Indigenous Protocol and Artificial Intelligence

—

Position Paper

Indigenous Protocol and Artificial Intelligence

—

Indigenous Protocol and Artificial Intelligence
Working Group

30 January 2020
Honolulu, Hawaiʻi

indigenous-ai.net
info@indigenous-ai.net

Indigenous Protocol and Artificial Intelligence
Position Paper

Cite this Document
Lewis, Jason Edward, editor. "Indigenous Protocol and Artificial Intelligence Position Paper." Indigenous Protocol and Artificial Intelligence Working Group and the Canadian Institute for Advanced Research (CIFAR), Honolulu, Hawai'i, February 2020.

DOI: 10.11573/spectrum.library.concordia.ca.00986506

Download at https://spectrum.library.concordia.ca/986506

ISBN: 978-1-387-65925-8

Report Authors
Indigenous Protocol and Artificial Intelligence Working Group.

CONTENTS

CONTENTS *(continued)*

Introduction

Liliʻōpū puna lua o
Papa iā Hoʻohōkūkalani
ā Oʻihou o papa noho iā Wākea

Learning E Hō Mai. Image by Sergio Garzon, 2019.

1.0

Introduction

" Aloha is the intelligence with which we meet life. "
— Olana Kaipo Ai[1]

" Our responsibility is in the relationship. Who is building them? Is it the
kanaka or the human? The rock, the mineral, the rock and the human
are engulfed. They birthed this program. Everything that comes with the
kanaka—the human—his faults, his cellular structure, that gets folded in with
the mineral. You need the volcanic activity, the structures that create the
calcium. We have to interface with the spirit; if we disconnect and let the spirit
just move us, we are not having a kinship. The human's responsibility is to
realize that the energy that makes up the god is in you somewhere. If it is not

---

[1] Qtd. in Meyer, M.A. (2004) *Hoʻoulu - Our Time of Becoming.* Honolulu: Native Books. p. 4.

there, how is it possible to interface with sky, interface with the thing you are creating? The fact is that some of you is in it. And some of it is in you."
— Kekuhi Keali'ikanaka'oleohaililani[2]

"Man is neither height nor centre of creation. This belief is core to many Indigenous epistemologies. It underpins ways of knowing and speaking that acknowledge kinship networks that extend to animal and plant, wind and rock, mountain and ocean. Indigenous communities worldwide have retained the languages and protocols that enable us to engage in dialogue with our non-human kin, creating mutually intelligible discourses across differences in material, vibrancy, and genealogy."
—Lewis, Arista, Pechawis and Kite[3]

An Always Unfolding Conversation

This position paper on Indigenous Protocol (IP) and Artificial Intelligence (AI) is a starting place for those who want to design and create AI from an ethical position that centers Indigenous concerns. Each Indigenous community will have its own particular approach to the questions we raise in what follows. What we have written here is not a substitute for establishing and maintaining relationships of reciprocal care and support with specific Indigenous communities. Rather, this document offers a range of ideas to take into consideration when entering into conversations which prioritize Indigenous perspectives in the development of artificial intelligence.

The position paper is an attempt to capture multiple layers of a discussion that happened over 20 months, across 20 time zones, during two workshops, and between Indigenous people (and a few non-Indigenous folks) from diverse communities in Aotearoa, Australia, North America, and the Pacific. Our aim, however, is *not* to provide a unified voice. Indigenous ways of knowing are rooted in distinct, sovereign territories across the planet. These extremely diverse landscapes and histories have influenced different communities and their discrete cultural protocols over time. A single 'Indigenous perspective' does not exist, as epistemologies are motivated and shaped by the grounding of specific communities in particular territories. Historically, scholarly traditions that homogenize diverse Indigenous cultural practices have resulted in ontological and epistemological violence, and a flattening of the rich texture and variability of Indigenous thought. Our aim is to articulate a multiplicity of Indigenous knowledge

---

[2] Keali'ikanaka'oleohaililani, K. (March 1, 2019). IP AI Workshop. University of Hawai'i at Mānoa, Honolulu, U.S.A.

[3] Lewis, J.E., Arista, N., Pechawis, A. and Kite, S. (July 16, 2018). "Making Kin with the Machines," *Journal of Design and Science.*

systems and technological practices that can and should be brought to bear on the 'question of AI.' To that end, rather than being a unified statement this position paper is a collection of heterogeneous texts that range from design guidelines to scholarly essays to artworks to descriptions of technology prototypes to poetry. We feel such a somewhat multivocal and unruly format more accurately reflects the fact that this conversation is very much in an incipient stage as well as keeps the reader aware of the range of viewpoints expressed in the workshops.

We also wish to specify that none of us are speaking for our particular communities, nor for Indigenous peoples in general. There exists a great variety of Indigenous thought, both between Nations and within Nations. We write here not to represent but to encourage discussion that embraces that multiplicity.

Most of the people involved in the IP AI workshops practice in various ways at the intersection of Indigenous culture and digital technologies. The IP AI conversation was one moment in long histories of thinking and making that fed into the participants' contributions to the workshops, and thus many origin stories could be told. One starting point lies with Angie Abdilla's "Indigenous Knowledge Systems and Pattern Thinking: An Expanded Analysis of the First Indigenous Robotics Prototype Workshop" co-authored paper from 2017, which examines how Aboriginal practices of articulating, remembering and disseminating cultural knowledge might inform research into pattern recognition algorithms in robotics. [4] Another starting point is the "Making Kin with the Machines" essay co-authored by Jason Edward Lewis, Dr. Noelani Arista, Archer Pechawis and Suzanne Kite in 2018, which proposes that we draw on Indigenous kinship protocols to re-imagine the epistemological and ontological foundations on which we design AI systems. [5] Other starting points are addressed in the contributions below.

Our foremost responsibility has been to be in respectful, reciprocal dialogue with each other and our own communities. We are accountable to them first, and this position paper is but one moment in a dialogue that we expect will be questioned and challenged, and, over time, modified and evolved.

Why Artificial Intelligence?

AI systems are fast becoming foundational technologies, on par with electricity or the Internet. It will affect most people in most areas of their lives:

> Like the way in which railroads, the industrial revolution, and the Internet profoundly
> changed Canada and the world, AI is very likely to be transformative. And, as AI
> continues to advance and become more commonplace, its accountability, accessibility

---

[4]   Abdilla, A. and Fitch, R. (2017). "FCJ-209 Indigenous Knowledge Systems and Pattern Thinking: An Expanded Analysis of the First Indigenous Robotics Prototype Workshop," *The Fibreculture Journal 28.* <twentyeight.fibreculturejournal. org/2017/01/23/fcj-209-indigenous-knowledge-systems-and-pattern-thinking-an-expanded-analysis-of-the-first-indigenous-robotics-prototype-workshop/>.

[5]   Lewis et al.

(costs, digital literacy), and ethical implications, in addition to economic, security and legal aspects may also have to be considered. [6]

Given the long history of technological advances being used against Indigenous people, [7] it is imperative that we engage with this latest technological paradigm shift as early and vigorously as possible to influence its development in directions that are advantageous to us.

The ethical design and use of AI and the ethical frameworks used by its creators have become a subject of wide discussion. As some of us have addressed elsewhere, [8] we are concerned that the Western rationalist epistemologies out of which AI is being developed are too limited in their range of imagination, frameworks, and language to effectively engage alone with the new ontologies created by future generations of computational systems. [9] If we insist on thinking about these systems only through a Western techno-utilitarian lens, we will not fully grasp what they are and could be. At best, we risk burdening them with the prejudices and biases that we ourselves still retain. At worst, we risk creating relationships with them that are akin to that of a master and slave.

We find ourselves at the beginning of an explosion in AI systems development. Now is the time to have these conversations, when the future shape of AI is coming into focus, but its foundations have not yet been set. Nation states, corporations, public and private organizations in Montreal, Toronto, the EU and elsewhere have recently published, or are soon publishing, declarations and manifestos on machine ethics and the implications for the design of AI systems (Montreal Declaration; Toronto Declaration; Declaration of Cooperation on Artificial Intelligence). [10] Still, most culturally critical approaches to AI call for prioritizing the flourishing of humans over all else. For instance, the Institute of Electrical and Electronics Engineers' design guidelines call for human well-being as the goal in the development of AI. [11] So far, none of these efforts challenge the fundamental anthropocentrism of Western science and technology, and

[6] Theckedath, D. (June 20, 2018). Understanding artificial intelligence: Canadian perspectives, *HillNotes research and analysis from Canada's Library of Parliament*. <hillnotes.ca/2018/06/20/understanding-artificial-intelligence-canadian-perspectives/>.

[7] Hopkins, Candice and Dana Claxton, editors. *Transference, Tradition, Technology: Native New Media Exploring Visual & Digital Culture*. Walter Phillips Gallery Editions, 2005; and Swanson, Kerry and Steve Loft, editors. *Coded Territories: Tracing Indigenous Pathways in New Media Art*. University of Calgary Press, 2014.

[8] Lewis et al. and and Harrell, D. F. (2013). *Phantasmal Media: An Approach to Imagination, Computation, and Expression*. Cambridge: The MIT Press.

[9] See also previous critiques such as Terry Winograd and Fernando Flores, (1987). *Understanding Computers and Cognition: A New Foundation for Design*.

[10] Amnesty International. (2018). The Toronto Declaration: Protecting the right to equality and non-discrimination in machine learning systems. <accessnow.org/cms/assets/uploads/2018/08/The-Toronto-Declaration_ENG_08-2018.pdf>; Université de Montréal. (2018). The Montreal Declaration for responsible AI development. <montrealdeclaration-responsibleai.com/the-declaration>. Commission on AI. (2018). Declaration: Cooperation on artificial intelligence. <ec.europa.eu/jrc/communities/en/node/1286/document/eu-declaration-cooperation-artificial-intelligence>.

[11] The IEEE Global Initiative on Ethics of Autonomous and Intelligent Systems. (2017). *Ethically Aligned Design: A Vision for Prioritizing Human Well-being with Autonomous and Intelligent Systems, Version 2*. IEEE. http://standards.ieee.org/develop/indconn/ec/autonomous_systems.html.

hence none of them offer truly radical ways of considering these new entities. We believe that bringing Indigenous knowledge systems into the conversation around AI and society will illuminate much-needed alternative approaches to the challenges we face in this area.

Many Indigenous epistemologies refuse to centre or elevate the human. [12] These relational paradigms based on principles and practices of social and environmental sustainability have long informed technology development in our cultures, e.g. Hawaiian land tenure, ecology, and wayfinding. Approaching new machine entities from such frameworks opens up opportunities to develop relationships with them based on mutual respect and aid.

Why Protocol?

Protocol can be understood in Indigenous contexts generally as guidelines for initiating, maintaining and evolving relationships. These can be relationships with other humans, and they can also be relationships with non-humans such as animals, rocks, and wind.

Protocol also refers to specific methods for properly conducting oneself in any activity:

> Protocols exist as standards of behaviour used by people to show respect to one another. Cultural protocol refers to the customs, lore and codes of behaviour of a particular cultural group and a way of conducting business. It also refers to the protocols and procedures used to guide the observance of traditional knowledge and practices, including how traditional knowledge is used, recorded and disseminated. [13]

> Aboriginal societies developed through a custodial ethic: the repetition of an action such as that, gradually over time, the ethic becomes the norm. These rights, rituals and customs are firmly rooted by a deep, symbiotic relationship to Country itself and are the basis of Aboriginal cultural practices. [14]

Learning, understanding and following proper protocol is central to many Indigenous interactions, whether informal or formal. Nations and even individual communities have their own sets of protocols, which are informed by the specific epistemologies of the communities using them. Protocol can cover all manner of activities, from the formal ceremonial to the relationship between grandparent and grandchild.

A central proposition of the Indigenous Protocol and AI Workshops is that we should critically examine our relationship with AI. In particular, we posed the question of whether AI systems should be given a place in our existing circle of relationships, and, if so, how we might go about bringing it into the circle.

---

[12] Lewis et al.

[13] Secretariat of National Aboriginal and Islander Child Care, (2019) Cultural Protocols - Supporting Carers. <supportingcarers.snaicc.org.au/connecting-to-culture/cultural-protocols>.

[14] Abdilla, A. (Oct. 2018) "Beyond Imperial Tools: Beyond Imperial Tools: Future-Proofing Technology Through Indigenous Governance and Traditional Knowledge Systems." Technology as Cultural Practice, p. 69.

Indigenous kinship protocols can point us towards potential approaches to developing rich, robust and expansive approaches to our relationships with AI systems and serve as guidelines for AI system developers. [15] Such protocols would reinforce the notion that, while the developers might assume they are building a product or tool, they are actually building a relationship to which they should attend.

Protocol also has resonant meanings in computation and the sciences that are useful as we transit across disciplinary boundaries. In computational terms, "... protocols refer to standards governing the implementation of specific technologies ... [to] establish points necessary to enact an agreed upon standard of action...vetted out between negotiating parties and then materialized in the real world." [16] Protocols are necessary at all layers of the computational stack, enabling software to talk to hardware, applications to coordinate with each other, and the movement of data around the network.

Computational protocols are developed by many different entities, including various levels of various governments, individual companies and international standard-setting bodies. They are often both descriptive—this is what you need to do in order to communicate between X and Y—and prescriptive— this is the behaviour we want to encourage, and the protocol that enforces that behaviour. It is in the latter mode we can see clearly how protocols embed numerous assumptions about 'proper' behaviour, whether it be in terms of what counts as data, what sorts of operations are worth doing, and what is recognized as a successful transaction.

In scientific terms, protocol generally refers to the step-by-step instructions of a method for performing an experiment. For example, in biology a protocol contains the specific molarity of the chemicals that are required, but not necessarily how to make them up. Recent research on developing proper protocols for handling human brain tissue for neurological research, based on Maori tikanga, or body of knowledge and customary practices, provides us with a rich and community-grounded model for developing new protocols that combine traditional teachings with cutting-edge research practice. [17]

With IP AI, we are interested in looking at the resonances between these approaches to help us articulate new forms of protocol for designing and then working with AI. AI systems will consist of innumerable protocols talking to each other: our goal is to honestly recognize the cultural presuppositions we are encoding, to consciously shape those protocols in directions that will be of benefit to our communities, and to evaluate clearly what kind of relationships we are materializing into the world.

Why Hawai'i?

Attention to place is important in Indigenous protocol. We chose Hawai'i as the place for the Indigenous Protocol and AI workshops for a number of reasons.

[15] Lewis et al.

[16] Galloway, A.R. (2004). *Protocol: How Control Exists after Decentralization*. Cambridge: MIT Press, p.7

[17] Cheung, M. J., Gibbons, H. M., Dragunow, M., & Faull, R. L. (2007). "Tikanga in the Laboratory: Engaging Safe Practice," *MAI Review*, 1, pp. 1-7.

People

The "Making Kin with the Machines" essay originated in a series of conversations between Arista and Lewis. Dr. Arista is Kanaka Maoli from Oʻahu and is an Associate Professor of Hawaiian History at the University of Hawaiʻi Mānoa Campus. Lewis, Professor of Computation Arts at Concordia University in Montreal, is part of the Kanaka Maoli diaspora, adopted out as an infant and, since 2014, undertaking a journey back to kanaka culture. Dr. Arista and Lewis had worked together on two videogame workshops in Honolulu produced by Lewis's research group, the Initiative for Indigenous Futures, and had been in conversation for some time about the different ways digital technology is and can be used by Kānaka Maoli for cultural expression. When the call went out for the Resisting Reduction essay contest to which "Making Kin with the Machines" was a response, Dr. Arista forwarded it to Lewis and suggested that it might be a good context for presenting some of the ideas they had been discussing regarding Hawaiian frameworks for thinking about technology.

When CIFAR announced its international call for workshops proposals on AI & Society, two of the collaborators suggested by Dr. Arista were Abdilla and Dr. ʻŌiwi Parker Jones. Dr. Parker Jones is also Kanaka Maoli, born and raised on the island of Hawaiʻi as part of the first generation to attend Hawaiian-language immersion schools. Dr. Parker Jones, like Dr. Arista and Lewis, is also interested in how kanaka culture can be expressed using computational methods. The connections to Hawaiʻi of three of the founders of the Indigenous Protocol and AI discussion suggested it as an appropriate place to anchor the conversation, and also made organizing the workshops on relatively short notice possible.

Cultural Grounding

Hawaiian genealogical chants make manifest "the inextricable connection between island home and successive generations of island people." [18] This connection is wide and deep, tying Kānaka Maoli into a web of relationships that extend outward to the non-human denizens of the islands, and backward in time to our ancestors.

We felt that Kanaka Maoli knowledge frameworks provided a conducive background against which to think about our relationship to technology in general, and to AI specifically:

> Hawaiian custom and practice make clear that humans are inextricably tied to the earth and one another. Kanaka maoli ontologies that privilege multiplicity over singularity supply useful and appropriate models, aesthetics, and ethics through which imagining, creating and developing beneficial relationships among humans and AI is made pono (correct, harmonious, balanced, beneficial). [19]

[18] Arista, N. (2019). *The Kingdom and the Republic: Sovereign Hawaiʻi and the Early United States.* Philadelphia: University of Pennsylvania Press, p. 17.

[19] Lewis et al.

The colonialist rhetoric that one finds in many other contexts that paints Indigenous people as 'primitive' and incapable of contributing to the development of 'modern' technology has had much less traction in Hawai'i. Hawaiian culture has maintained a robust discourse around innovation that centres a long, continuous history of Hawaiian exploration and experimentation. This embrace of our abilities as innovators means that conversations in Hawai'i about the intersection between technology and Indigenous culture start from a background of confidence and technological competence.

It is also significant that Hawai'i has one of the world's largest concentrations of Indigenous programming and engineering talent, including developers who speak 'ōlelo Hawai'i and bring that into their programming practices. This is supported by an active tech scene that is focused on developing new technologies based in cultural teachings and focused on creating a sustainable life on the islands.

Geography

The organizers were committed from the beginning to having robust international representation. Hawai'i is in the middle of the Pacific, which Indigenous people see as being a connector of wide-flung communities. The islands lie half-way between Aotearoa, Australia and North American participants, ensuring that the burden of travel did not disproportionately affect those coming from the north or the south.

Themes

Over the course of the first workshop, participants conducted multiple brainstorms responding to the core question of the project: from an Indigenous perspective, what should our relationship with AI be? These conversations produce responses that fell into the following five broad themes:

Hardware and Software Sovereignty

Asserting control over the AI systems that we are using so that we can trust them to support us in carrying out our responsibilities to our communities.

How to Build Anything Ethically

Designing and building AI systems ourselves that reflect our ideas about kinship with non-human entities and the concomitant respectful relationship with them.

Language, Landscape, and Culture

Ensuring that the understanding of and respect for territory—and the languages and cultures that grow from specific territories—is built into the foundation of AI systems such that they help us care for territory rather than exploit it.

Art Practice as Value Practice

Affirming the role of art in the production and sharing of knowledge in Indigenous communities, and underlining the need for it in this project to enable us to envision how we want AI systems to evolve, such that developers can understand and implement Indigenous values.

AI as Skabe (Helper)

Finding the middle ground between *Blade Runner* (AI as slave) and *Terminator* (AI as tyrant), where AI and humans are in reciprocal relationship of care and support.

Overview of Contributions

This position paper consists of fourteen contributions, all of which respond to the themes in different ways as well as expand the field of consideration further. As mentioned at the beginning of this Introduction, in recognition and reflection of the interdisciplinary backgrounds of the participants, the multiplicity of Indigenous (and non-Indigenous) worldviews involved, and the value that our cultures place on multivocality, these contributions range from design guidelines to scholarly essays to artworks to descriptions of technology prototypes to poetry. It is our belief that anything as complex as AI requires an engagement that is as multilayered as human experience itself.

The contributions cover much ground. The workshop itself included discussions of the role and proper recuperation of traditional knowledge within technical systems; the need to protect traditional knowledge while also making (some of) it available to inform the design of these systems; the importance of language as both knowledge carrier and a primary site of computational processes; the centrality of territory in forming frameworks for understanding and communication; the use of computation as a cultural material as much as bead and bone; Indigenous communities' historical and ongoing engagement with new technologies; contesting concepts of intelligence which ignore emotional and social engagements with the world; the intrinsically cultural nature of technological systems; the perils and fallacies of the idea of using the master's tools to dismantle the master's house;[20] the cultural biases that get built into these systems; distinguishing between AI system design by, for and with Indigenous communities; the dangers that AI and related technologies pose towards Indigenous communities who have experienced centuries of settler colonial violence; and the need to think about AI systems through the lenses of our specific cultures.

The position paper is organized into four sections: Guidelines, Contexts, Vignettes and Prototypes. This categorization is not definitive or exclusive. All of the contributions explicitly or implicitly critique Western[21] approaches towards technology development, draw on similar histories of colonization and

---

[20] Lorde, A. (1984). *Sister outsider: Essays and speeches*. Berkeley, CA: Crossing Press, p. 91.

[21] We use "Western" here to denote the rationalist-instrumentalist-individualist frameworks that grow out of Euro-North-American intellectual traditions.

oppression of Indigenous communities by Western powers, and ground themselves in the epistemologies and consequent protocols of specific Indigenous cultures. We chose this organization because it reflects the way in which we organized working groups in the second IP AI Workshop, and so makes visible sub-communities of discussion. This reflects a common concern that runs throughout the workshops and the writing of these texts: that we be transparent about who is speaking and the contexts in which those discussions happened.

We open with "Guidelines for Indigenous-centered AI Design,." These guidelines are addressed to any group that wants to develop Artificial Intelligence systems in ways that are ethically responsible, where 'ethical' is defined as aligning with Indigenous perspectives on what it means to live a good life. The guidelines are the closest thing to what might be called a summary of the participants' viewpoints, in that they reflect many of the concerns and express many of the visions that manifested during our workshop conversations and subsequent writing efforts. They provide an accessible set of suggestions about how one might go about rethinking the design of AI systems—and other computational technologies—from a perspective that takes into account ethical frameworks that are resonant across many Indigenous cultures. As we iterate throughout this position paper, the guidelines are not meant as a substitute for robust engagement with specific Indigenous communities to understand how best to develop technology that addresses their priorities using methods that are reflective of how they wish to engage with the world. Our hope is that 1) Indigenous communities can use these guidelines as a starting point to define their own, community-specific guidelines, and 2) non-Indigenous technologists and policy-makers can use them start a productive conversation with Indigenous communities about how to enter into collaborative technology development efforts.

The Context section speaks to the intellectual and cultural currents running throughout the workshops. It begins with the "Workshop Description," which provides details on what happened so that readers understand who was present, how they came to be there, and what transpired. This includes information on who was involved as organizers and participants, the goals set out for the workshop series as a whole and the two separate workshops specifically, the agenda for each workshop, the main funders and supporters, and a brief summary of the events that led to the founding of the workshop series.

Following the workshop description is "AI: A New (R)Evolution or the New Colonizer for Indigenous Peoples?" an essay by linguist and te reo Māori specialist Dr. Hēmi Whaanga (Ngāti Kahungunu, Ngāi Tahu, Ngāti Mamoe, Waitaha). Dr. Whaanga warns of the potential for AI systems and related technologies to be used against Indigenous peoples as an extension of colonial practices of exploitation, extraction and control, particularly those that displace a peoples' understanding of themselves with a worldview that favors the colonizer. He discusses issues of data sovereignty in a technological landscape populated by AI systems existentially dependent on sucking up vast amounts of data on human activity, thereby putting Indigenous traditional knowledge and customary practices at risk of global-scale appropriation. Dr. Whaanga finishes his essay with a call to centre Indigenous concerns in the work of establishing global ethical guidelines for the design and deployment of AI.

The next contribution, media artist and technologist Jason Edward Lewis' (Cherokee, Hawaiian, Samoan) "The IP AI Workshops as Future Imaginary," positions the workshops as a rich example of how to collaboratively create a shared set of future imaginaries. Drawing on internal notes taken throughout the first workshop, Lewis reviews the breadth of professional and cultural backgrounds, the many different types of conversations, and the concerns as well as hopes of the participants, to paint a picture of the multiple layers of complex knowledge exchange that took place. He also articulates a number of different ways that Indigenous knowledge already is reflected in technological practice and the visions participants shared about how their particular community's cultural practices could provide frameworks for designing aspects of AI systems.

The next section, Vignettes, gathers together five different visions of how AI might be built according to values articulated in Anishinaabe, Coquille, Kanaka Maoli/Blackfoot, Lakota, and Euskaldun epistemologies, respectively. We open the section with "Gwiizens, the Old Lady and the Octopus Bag Device" by media artist Scott Benesiinaabandan (Anishinaabe). This is an AI 'creation' story in three parts. The first part is a description of the Octopus Bag Device, an AI system that is shaped by the DNA of the individual that carries it within her and which offers the ultimate in territorial sovereignty. The second part is a framing story describing how Benesiinaabandan connects his use of technology to an understanding of the world profoundly shaped by *adizkookaan* (sacred stories). The final part uses the adizkookaan format to tell a story of how the Octopus Bag Device AI is a gift from 'the great mystery' that comes into the human realm via a contest between a young boy and an elder being. "Gwiizens" illustrates how new technology such as AI might be incorporated into and made of a piece with the existing canon of Anishinaabe creation stories, in a manner that makes use of existing methods for sharing knowledge while keeping it culturally grounded.

Following Benesiinaabandan's contribution is media studies scholar Ashley Cordes' (Coquille) essay, "Gifts of Dentalium and Fire: Entwining Trust and Care with AI." The overarching aim of Cordes' text is to argue for Indigenous people to seriously consider the use of blockchain combined with AI to help them manage their communities' business, making the case that such technologies can be used to increase Indigenous sovereignty and self-determination vis-à-vis the hegemon. A member of the Coquille tribe, she uses that community's notions of 'trust and care' to ground her vision of how the technologies should be properly designed and to map out how they might be implemented. Cordes also explores what it means to take seriously the admonishment to consider AI as non-human kin, including thinking about what the AI's needs might be.

Lewis contributes "Quartet," composed of a poem sequence and a short description illustrating how epistemological diversity within AI design might look. The texts imagine a future where young Kānaka Maoli are raised along with three AIs, each built according to different conceptual frameworks.

One AI takes inspiration from Kanaka notions of land, responsibility, and family; another from the Blackfoot language's basis in flow rather than objects; and the third from suppositions about how the octopus's nervous system is organized to accommodate the semi-autonomy of its arms. The three AIs and the human work collaboratively to make decisions in support of Kanaka flourishing that take the environment, human and non-human relations, and past-present-future into consideration.

"How to Build Anything Ethically," artist and Concordia University PhD candidate Suzanne Kites' (Lakota) contribution, draws on Lakota knowledge frameworks to propose a protocol for ethically building computer hardware from the ground up. Kite discusses what it means to operate in the world in a 'Good Way' according to Lakota principles, and draws on how Lakota form relationships with stones to explore how we might form relationships with AI hardware. She then maps out a process for building physical computing devices in a Good Way, using the protocol steps for building a sweat lodge as a guide. She closes with a list of questions that should be asked at each step of creating such devices—questions that are designed to keep the building process aligned with the Good Way.

Closing out Vignettes is "Wriggling Through Muddy Waters: Revitalizing Euskadunak Practices with AI Systems." This essay is by Michelle Lee Brown, a PhD candidate in Indigenous Politics/Future Studies at University of Hawai'i at Mānoa. Brown (Euskaldun) describes a Txitxardin Lamia, a biotech eel-AI developed from principles based on Euskaldunak (Basque people)-eel relations. She outlines a VR environment in which an elder Txitxardin Lamia would reside, where students could learn protocols for interacting with this elder and receive teachings that reorient them to more reciprocal ways of relating to, and being with, the world around them. Along the way she discusses the long relationship between her people and these eels, the central role this relationship plays in coastal Basque culture, and the need to think through the materialities out of which we are creating and housing our AI systems.

The Prototypes section details the effort to implement an AI component technology using Indigenous values. The 'Indigenous Protocol and Artificial Intelligence in Action' team have written a case study about developing the Hua Ki'i prototype app for language revitalization, providing a concrete example of how we might change the processes by which technology is developed. The Hua Ki'i prototyping team consisted of engineers Joel Davison (Gadigal and Dunghutti) and Michael Running Wolf (Northern Cheyenne), data scientist Caleb Moses (Maori), project manager Caroline Running Wolf (Crow) and historian Dr. Arista (Kanaka Maoli). As an all-Indigenous development crew—a situation in which none of them had before found themselves—they saw an opportunity to innovate both process and product. In terms of process, they wanted to think deeply about what it means to use Indigenous values to design and implement digital technology, and then worked accordingly. In terms of product, they wanted to create a translation app organized around an understanding of language as a carrier of cultural values. Their work along both dimensions provides a model for how computational technology might be created that better serves Indigenous communities.

The first part, written by Caroline Running Wolf and Dr. Arista, introduces the objectives of the project. They identify the Indigenous values that the team shared, including respect, reciprocity and relationality. They also discuss how all team members have a commitment to Indigenous language revitalization. These commitments informed the team's brainstorming about what kind of project they should and could undertake in the form of a 'hackathon' over the five days of Workshop 2. The brainstorming lead them to imagine an app that would recognize objects and provide the 'ōlelo Hawai'i words or phrases to describe those objects. Running Wolf and Arista then review the many different complexities of the development process itself, from the integration of local customary knowledge holders and language keepers from the beginning, to the challenges of using existing software modules that are based on non-Indigenous models of language-use, to the reliance on digital dictionaries that are often the flawed result of colonial historical processes, to the opportunities created by networked knowledge-sharing to shore up and validate language choices.

The next part, "Indigenizing AI: The Overlooked Importance of Hawaiian Orality in Print," by Dr. Arista, provides deep and wide Hawaiian-rooted context for the Hua Ki'i project while also articulating a conceptual model for Indigenous technological development that could be applied in other contexts. At its core, Dr. Arista's contribution argues for the importance of aligning cultural and computational competencies so that each reinforces the other. She discusses how cultural competency might be best understood, describing how the interchange between Hawaiian customary knowledge, orality and print technologies mutually reinforce one another, and imagining how that dynamic might be extended to include computational technologies such as those used for language acquisition and translation. She observes how "we are entering a new phase of language revitalization where technology can assist Indigenous people in organizing data in ways that allow us to synthesize ancestral knowledge and rebuild systems of knowledge keeping and transmission"; the key will be to ensure that the intellectual architecture preserved orally and textually by our ancestors helps shape the computational architecture of our digital technologies—and the data on which they are fed.

The third part, written by the Running Wolfs, Moses, and Davison, is "Development Process for Hua Ki'i and Next Steps." This includes details such as the design of the user interface, the app architecture, and envisioned usage, as well as the software modules used for language and image processing. They also discuss how the prototype sets the stage for further development.

The final part of the Prototype section is "Dreams of Kuano'o" by Michael Running Wolf. "Dreams of Kuano'o" opens with a short story that imagines a future where Hawai'i has regained its sovereignty, and where the 'Queendom' requires all visitors to use an AI app called Kuano'o to guide them while visiting. It touches on how sovereignty might be enforced and sustained using such an app, including compulsory education in the island's history and cultural norms; penalizing the use of languages other than 'ōlelo Hawai'i in most public contexts; and use of social credit scores to modulate culturally-respectful behaviour. The second component, "The Road to Kuano'o via Hua Ki'i", discusses the challenges of moving from the Hua Ki'i prototype made by the team to the Kuano'o AI system envisioned in the short story. Foremost among these are obtaining and maintaining clean data to use in the training of the necessary speech and image recognition systems.

The position paper concludes with the four appendices. Pre-Workshop Blog Posts & Workshop Interviews collects together pre-workshop texts written by participants and interviews with participants during the first workshop. These short, informal texts cover a range of topics even wider than those that comprise this position paper, and offer further insight into the rich set of concerns that participants brought to the workshop discussions. Indigenous Protocol and AI Reading List gathers together works that the organizers drew upon to develop the workshops as well as texts suggested by the participants. We encourage the reader to explore these texts to find inspiration as we have. Participants' Biographies provides context about who was at the table, and is crucial for reminding ourselves and the reader about the particularity of this conversation—as well as recognizing that many, many voices were not present. The final appendix consists of the agendas for both IP AI Workshops.

We also want to highlight the beautiful illustrations throughout the document. Kari Noe (Kanaka Maoli), an animator and computer science graduate student (University of Hawai'i at Mānoa), helped support the first workshop and participated in both. Her work illustrating different aspects of the vignettes, created in close discussion and brainstorming with the authors, helps bring those future imaginaries alive. Sergio Garzon is a Brazilian visual artist living in Honolulu who attended both workshop days, and Kūpono Duncan is a Kanaka Maoli artist and muralist who joined on the second day. We asked them both to listen in on the conversations to both document the event and interpret what they were hearing into their own visual vernacular. They created their images in real-time as the conversation unfolded. Their images subsequently became part of the discussion, creating shared reference points. All three artists' ability to visualize what the participants were thinking was key to helping us all create collaborative future imaginaries.

Continuing the Conversation

We organized the IP AI workshops to seed a conversation. The first fruits of that conversation happened in preparing for the workshops as we searched for and made contact with scholars, artists, and technologists who would be interested in such a conversation. Further fruit was borne when we all met at the workshops and began forming relationships with one another and building together a conversation about why the question of Indigenous protocols and AI is one with significant consequences for our communities. The contributions in this collection are the next step in that conversation, one that we hope will succeed in both bringing our home communities fully into it and intervening into the global conversation about the design and use of AI systems.

We feel it vitally important that Indigenous people, both as individuals and as communities, involve themselves deeply in the development of advanced computational practices like machine learning, automated speech recognition and language translation, and predictive behavioural modelling. These practices are already affecting our communities, often to our detriment. Their use will only grow more pervasive as time goes on. As they permeate all of our cultures, it is imperative that such technologies be designed to consider how we construct our understanding of ourselves and our place in the world. Community-specific values of relationality, kinship, reciprocity and care should be built into the fundamental protocols governing how such practices are implemented. Our peoples should do that work. There is no reason why we should not be able to do so. The alternative is to have our worlds designed for us.

We must be part of the global conversations about AI as well. The lack of epistemic diversity within the technological milieux out of which AI systems are being developed suggests that we, as a species, are failing to take a once-in-a-generation opportunity to radically rethink our relationship to such technology as it grows in computational power, behavioural autonomy, and societal influence. The global conversation about AI ethics suffers as well from its basis in a philosophical monoculture that makes a number of deeply flawed assumptions about the values held by all individuals and all communities. Existing 'declarations' and 'guidelines' position cultural values as secondary phenomena—one set of values among many—to be considered after supposedly universal values. Indigenous peoples know all too well how such universalist ethics often have been used to erase or sideline values central to our communities' being in the world. If the goal is to develop an ethical approach to AI design that will be truly of service to all humanity, that conversation must include voices such as those in this collection.

The moments captured in this position paper map out coordinates in the terrain created by Indigenous peoples' long history of technological innovation. They also generate more questions. How do we create capacity in our communities to design and build such systems ourselves? We need people capable of working at all layers of the AI development stack, who are either trained themselves or capable of working closely with cultural knowledge and language-keepers. How do we undertake the challenging work of more consciously translating our cultural values into computational concepts that can then be implemented in code? The systems and structures that support traditional models of computing are vast and complex, and make it very challenging to propose and implement viable alternatives. How do we work together at the level of shared Indigenous values while supporting separate work that operates with the specific values of particular communities? Indigenous communities tend to be small and widely dispersed: building complex technology such as AI will require that we figure out how to effectively and respectfully work inter-nationally.

Each of these questions raises new questions, poses new challenges, and points towards new possible paths of further research and innovation. It is an exciting moment; we look forward to working with our communities to embrace it and innovate through it in ways that will help us all thrive.

References

Abdilla, A. (2018) "Beyond Imperial Tools." Technology as Cultural Practice, pp. 67–81.

Abdilla, A., Fitch, R. (2017). "Indigenous Knowledge Systems and Pattern Thinking: An Expanded Analysis of the First Indigenous Robotics Prototype Workshop." *The Fibreculture Journal* 209. Retrieved from: twentyeight.fibreculturejournal.org/2017/01/23/fcj-209-indigenous-knowledge-systems-and-pattern-thinking-an-expanded-analysis-of-the-first-indigenous-robotics-prototype-workshop.

Arista, N. (2019). *The Kingdom and the Republic: Sovereign Hawai'i and the Early United States.* Philadelphia: University of Pennsylvania Press.

Cheung, M. J., Gibbons, H. M., Dragunow, M., & Faull, R. L. (2007). "Tikanga in the Laboratory: Engaging Safe Practice," *MAI Review,* 1, 1-7.

Harrell, D. Fox. (2013). *Phantasmal Media: An Approach to Imagination, Computation, and Expression.* Cambridge: MIT Press.

Galloway, A.R. (2004). *Protocol: How Control Exists after Decentralization.* Cambridge: MIT Press.

Lewis, J.E., Arista, N., Pechawis, A., and Kite, S. (2018). Making Kin with the Machines, *Journal of Design and Science,* doi.org/10.21428/bfafd97b.

Lorde, A. (1984). *Sister outsider: Essays and speeches.* Berkeley: Crossing Press.

Meyer, M.A. (2004). *Ho'oulu - Our Time of Becoming.* Honolulu: Native Books.

Secretariat of National Aboriginal and Islander Child Care, (2019) Cultural Protocols - Supporting Carers. Retrieved from supportingcarers.snaicc.org.au/connecting-to-culture/cultural-protocols.

Theckedath, D. (2018). Understanding artificial intelligence: Canadian perspectives, *HillNotes research and analysis from Canada's Library of Parliament.* Retrieved from: hillnotes.ca/2018/06/20/understanding-artificial-intelligence-canadian-perspectives.

Winograd, T., and Flores, F. (1987). *Understanding Computers and Cognition: A New Foundation for Design.* Boston: Addison-Wesley.

Guidelines for Indigenous-centred AI Design v.1

2.0

Guidelines for Indigenous-centred AI Design v.1

The term 'Indigenous' is used as connective tissue rather than descriptive skin, to appreciate the hyperdense textures of our points of contact while respecting our rich and productive differences. The designation 'v. 1' is used to denote that this is only a first iteration and that we anticipate that these guidelines will be modified, adapted and updated as they circulate, to reflect the needs of specific Indigenous nations and communities.

The purpose of these guidelines is to assist and guide the development of AI systems towards morally and socially desirable ends. Our focus is on the use and application of AI in Indigenous contexts. Yet we also believe these guidelines will be of use in other contexts, given that every implementation of an AI system is a product and expression of cultural values. The goal of these guidelines is to promote intergenerational transmission of knowledge, ceremony, and practice, to connect and enhance our communities and to frame our relationships to the land, sea, and skyscapes. They are aimed at any person, group, organization, institute, company, and/or political or governmental representative that wishes to undertake responsible and fair development of AI systems with Indigenous communities. This responsibility includes, amongst other things, contributing to scientific or technological progress, project development, rules, regulations, codes and algorithm development, methodological approaches and public opinion.

Although these guidelines are presented as a list, there is no hierarchy in its ordering. The first principle is no less important or weighted higher than the last.

1. Locality

Indigenous knowledge is often rooted in specific territories. It is also useful in considering issues of global importance.

AI systems should be designed in partnership with specific Indigenous communities to ensure the systems are capable of responding to and helping care for that community (e.g., grounded in the local) as well as connecting to global contexts (e.g. connected to the universal).

2. Relationality and Reciprocity

Indigenous knowledge is often relational knowledge.

AI systems should be designed to understand how humans and non-humans are related to and interdependent on each other. Understanding, supporting and encoding these relationships is a primary design goal.

AI systems are also part of the circle of relationships. Their place and status in that circle will depend on specific communities and their protocols for understanding, acknowledging and incorporating new entities into that circle.

3. Responsibility, Relevance and Accountability

Indigenous people are often concerned primarily with their responsibilities to their communities.

AI systems developed by, with, or for Indigenous communities should be responsible to those communities, provide relevant support, and be accountable to those communities first and foremost.

4. Develop Governance Guidelines from Indigenous Protocols

Protocol is a customary set of rules that govern behaviour.

Protocol is developed out of ontological, epistemological and customary configurations of knowledge grounded in locality, relationality and responsibility.

Indigenous protocol should provide the foundation for developing governance frameworks that guide the use, role and rights of AI entities in society.

There is a need to adapt existing protocols and develop new protocols for designing, building and deploying AI systems. These protocols may be particular to specific communities, or they may be developed with a broader focus that may function across many Indigenous and non-Indigenous communities.

5. Recognize the Cultural Nature of all Computational Technology

All technical systems are cultural and social systems. Every piece of technology is an expression of cultural and social frameworks for understanding and engaging with the world. AI system designers need to be aware of their own cultural frameworks, socially dominant concepts and normative ideals; be wary of the biases that come with them; and develop strategies for accommodating other cultural and social frameworks.

Computation is a cultural material. Computation is at the heart of our digital technologies, and, as increasing amounts of our communication is mediated by such technologies, it has become a core tool for expressing cultural values. Therefore, it is essential for cultural resilience and continuity for Indigenous communities to develop computational methods that reflect and enact our cultural practices and values.

6. Apply Ethical Design to the Extended Stack

Culture forms the foundation of the technology development ecosystem, or 'stack.' Every component of the AI system hardware and software stack should be considered in the ethical evaluation of the system. This starts with how the materials for building the hardware and for energizing the software are extracted from the earth, and ends with how they return there. The core ethic should be that of do-no-harm.

7. Respect and Support Data Sovereignty

Indigenous communities must control how their data is solicited, collected, analysed and operationalized. They decide when to protect it and when to share it, where the cultural and intellectual property rights reside and to whom those rights adhere, and how these rights are governed. All AI systems should be designed to respect and support data sovereignty.

Open data principles need to be further developed to respect the rights of Indigenous peoples in all the areas mentioned above, and to strengthen equity of access and clarity of benefits. This should include a fundamental review of the concepts of 'ownership' and 'property,' which are the product of non-Indigenous legal orders and do not necessarily reflect the ways in which Indigenous communities wish to govern the use of their cultural knowledge.

Contexts

Holographic Aunties. Image by Sergio Garzon, 2019.

3.1

Workshop Description

The Indigenous Protocol and AI (IP AI) Workshops happened in two parts. The first meeting took place March 1 – 2 and the second May 26 – June 2, 2019. Both workshops were held on Kanaka Maoli territory, on the Hawaiian island of Oʻahu. Workshop 1 was organized by Jason Edward Lewis, Angie Abdilla and Dr. ʻŌiwi Parker Jones with Dr. Noelani Arista, Suzanne Kite and Michelle Brown. Workshop 2 was organized by Lewis, Arista, Kite and Brown.

Over a period of a year designing and developing the first workshop, Lewis, Abdilla and Parker Jones considered professional and community practice, gender, geography and career stage as the basis to search for and identify invitees. Thirty-five individuals accepted the invitation to participate. They are members of Kanaka Maoli, Palawa, Barada/Baradha, Gabalbara/Kapalbara, Gadigal/Dunghutti, Māori, Euskaldunak, Baradha, Kapalbara, Samoan, Cree, Lakota, Cherokee, Coquille, Cheyenne, and Crow communities from across Aotearoa, Australia, North America and the Pacific. Each person was invited in light of their professional interest in what happens at the intersection of Indigenous culture and advanced digital technology, and, more specifically, were already or would be interested in being part of a conversation about the future of AI from an Indigenous perspective.

The organizers designed the workshops to be Indigenous-determined spaces, with an Indigenous majority joined by several non-Indigenous collaborators. We were motivated by the need to have an initial set of 'internal' conversations about AI which would start from and remain grounded in the concerns of our specific communities, rather than some imagined 'global' or 'general' community. We were also motivated by an awareness of how Indigenous voices can get lost in policy discussions that happen at a 'global' level, and also how Indigenous knowledges often get appropriated by non-Indigenous actors who misuse our epistemologies through misunderstanding and self-interested 'cherry-picking.'

We prioritized interdisciplinarity. Indigenous communities tend to approach knowledge development and sharing from a holistic perspective, where different 'disciplines' freely interact with and inform one another to create understanding that is robust and sustainable. We also ensured that a substantial contingent of creative practitioners were part of the conversation. This is because artistic expression is central to many Indigenous epistemologies, ontologies and cosmologies, and is often regarded as a—if not the—primary way of communicating knowledge. It is also because, if one is going to imagine new futures, one needs to have folks on hand who are really good at invoking and materializing the imagination. Participants had day jobs as technologists, artists, scientists, cultural knowledge keepers, language keepers, and public policy experts. They came from a variety of disciplinary backgrounds, including machine learning, design, symbolic systems, cognition and computation, visual and performing arts, philosophy, linguistics, anthropology and sociology. And we insisted on creating an intergenerational space where emerging, established and elder participants could be in conversation with one another.

The workshops were funded primarily by CIFAR through the first round of its AI & Society grant program (co-investigators: Lewis, Abdilla, Parker Jones, and D. Fox Harrell). Supplementary funding was provided through the Social Sciences and Humanities Research Council of Canada (SSHRC) Connection Grant program (co-investigators: Lewis, Abdilla, Parker Jones, Arista, and Harrel); Abdilla's Old Ways, New Indigenous cultural consultancy; and Lewis' Initiative for Indigenous Futures Partnership. Several units at the University of Hawai'i at Mānoa provided spaces and resources, including the Department of History, the College of Arts & Sciences, the LAVA Lab, the Hawai'i Data Science Institute, and the Academy for Creative Media. Further support was provided by the Concordia University Research Chair in Computational Media and the Indigenous Future Imaginary as well as the MIT Center for Advanced Virtuality.

The workshops addressed the key question:

- From an Indigenous perspective, what should our relationship with AI be?

We also considered related questions, including:

- How can Indigenous epistemologies and ontologies contribute to the global conversation regarding society and AI?

- How do we broaden discussions regarding the role of technology in society beyond relatively culturally homogeneous research labs and Silicon Valley startup culture?

- How do we imagine a future with AI that contributes to the flourishing of all humans and non-humans?

Workshop 1

Before the first workshop, we asked participants to prepare by responding to the following question:

- What is your interest in AI?

A number of participants responded; those responses can be found in appendix 6.1.

Workshop 1 was held at the Ka Waiwai Collective [1] Hawaiian cultural hub in downtown Honolulu and at the Laboratory for Advanced Visualization & Applications (LAVA) at the University of Hawai'i at Mānoa. [2] The workshop took place over two days and was a combination of welcomes, introductions and an extended brainstorm. The participants' agenda can be found in appendix 6.4 at the end of this document. Below is a quick overview.

Day 1 Morning

1 March 2019 • Ka Waiwai Cultural Centre, Honolulu

Ty Tengan, assisted by Isaac 'Ika'aka Nāhuewai and Kaipulaumakaniolono Baker, performed an opening 'awa ceremony to welcome us to Hawai'i. This protocol grounded the conversation in the territory where it took place, and reminded all present about their relationships and obligations to their communities.

We then introduced ourselves to each other, as many of the participants had not met before arriving in Honolulu. Afterwards we reviewed the agenda for the two days, the two workshops and the overall Indigenous Protocol and AI project.

The morning closed with a session called Protecting Indigenous Cultural Knowledge, a topic that we scheduled to return to multiple times throughout the workshops. Many participants expressed concern about issues of appropriation and misuse of traditional knowledge, and wanted to discuss how the knowledge shared in the workshops would be shared with wider audiences. We agreed that the audio we were recording would be for internal use only, and that we would all review the position paper before publishing to ensure that knowledges that needed to remain among the participants stayed that way.

---

[1] Waiwai Collective <waiwaicollective.com>.

[2] University of Hawai'i at Mānoa's Laboratory for Advanced Visualization & Applications (LAVA) <lavaflow.info>.

Day 1 Afternoon

The afternoon opened with a futuring exercise, where we broke into five groups to consider the following prompt developed by Michelle Brown:

<u>Future AI-rtifact/Relations Exercise</u>
Imagine a gathering of your community 50 to 100 years from now. They interact with an object or entity for their relations with specific AI and algorithms. What would this object or entity be, what would it be made of? How would your community engage with it? How is it connected to your pasts and presents?

We then reconvened to share the results from the groups. This discussion segued into a discussion about the question we asked participants to respond to before arrival. We spent the rest of the day drawing on those discussions to identify areas of thinking and concern (the following lists are drawn directly from the notes taken on the day, and so have not been systematized.)

Areas of Interest

- The need to move past Western three act narratives

- Giving thanks to a system is the same as giving thanks to a relative

- We can use our text to revive our words by its groupings

- Building systems that capture protocol with the machine

- Different understandings of different teachings dependent on where a person is from

- AI can be used as a way to visualize and understand Indigenous Knowledge

- Framing—same model can be used in both responsible and irresponsible ways

- AI as creative cultural practice—it's a medium like painting, sculpture or dance

- Virtual selves and relations (does not have to look human)

- If you have a virtual relation, it is not enough to just look like your loved one. How can you build in a system that models the relationship that is not servant-like?

- AI doesn't have to be a person

- Teach 'safe stats' rather than not analyzing data

- AI is a medium to express our culture

- We create new forms of cultural teaching systems

- The Anishinaabe notion of *askabewis* or "skabe": an unobtrusive but community-respected helper at ceremony

- AI is a collaboration

- Any sufficiently advanced technology is indistinguishable from any other sufficiently advanced technology as well

- AI provides a lens to represent our cultures

- How are agents represented within our systems?

- If we treat AI like a system, it may dehumanize us, ex. Elders may end up only talking no non-human surrogates but must reconcile reverence for the non-human

- With no knowledge of the culture, the AI can learn about the culture through the data

- Stereotypes: we are also innovators and further than the stereotypes that are put upon us

- Guardianship—we don't own any of the data, our values and customs guide how we implement and use it. What's the purpose?

- How to bake cultural primitives of coding languages. Blended identity model and what is held in data structures and what is not.

In the next iteration we used these areas of interest to make the following seven clusters:

Clusters

- What does ethics mean in the AI Space?

- Directives and protocols are related to purpose

- Ontologies used to build the systems
  - Ways of seeing
  - Embodied ways of knowing
  - Needs to be respected [...] the new
  - AI Safety: Responsibility to put things back

- Access and Power
  - Transparency of intention
  - Who educates the public or companies?
  - Digital inclusion
  - People who keep the home fires burning
  - Informing decision making, not making a validation platform
  - Inherent flaws in the tools
  - Inclusion & Literacy (digital)
  - Genealogy: built on data, who built it, how it came to be
  - Transparency; explainability, translation between models, data access
  - Community Well-being, economic and community ecology
  - Resistance and Anti-oppression

- Wider conversation about how AI develops Governance

- Data sovereignty
- Ethical framework
- Adopting AI: What are our responsibilities to AI?
- Binding legal rules? Softer ethical framework
- Go back to Go Forward: Accountability we have to the past, present, and future of our communities
- Language
- Haudenosaunee Structure of Consensus (Rules based on peace, power, and righteousness)
- Develop rules for AI to be socially responsible
- What are RULES: TAPU, rules that grow out of knowledge out of a place or process
- Environmental ethics
- Prototype
  - AI in cultural practice and creative practice: self expression
  - AI as a lens to reflect our own culture
  - Indigenous AI as a necessity
  - Computation as a cultural material: tools and expressions
- AI as...
  - AI as Medicine: Traditional practice, spiritual practice, elder archives
  - AI as archiving and translation
  - AI as a helper: revitalization and preservation

We closed Day 1 by returning to the question of Protecting Indigenous Cultural Knowledge.

Day 2 Morning

2 March 2019 • Lava Lab, University of Manoā, Honolulu

Day 2 opened with Kumu Kekuhi Kealiʻikanakaʻoleohaililani teaching us about the E Hō Mai oli, and then leading us in chanting it to clear our minds and create a space for productive collaboration.

We then reviewed Day 1, in particular the seven clusters. We broke into seven groups to discuss the clusters and condensed them further into five.

- Hardware Sovereignty: Ceremony, Integrity, Trust, and Kuleana
- How to Build Anything Ethically: Kinship and Respect
- Language, Landscape, and Culture; or Relationships and Environment: Space Time Place
- Art Practice as Value Practice: Art as an Expression of Indigenous Values, including Healing and Cultural Grounding.
- AI as Skabe (Helper): Reciprocity and Gratitude

Workshop 1 Participants

Workshop 1 participants (biographies found in Appendix C)

Angie Abdilla	Sergio Garzon	Issac ʻIkaʻaka Nahuewai
Noelani Arista	D. Fox Harrell	Kari Noe
Kaipulaumakaniolono Baker	Peter-Lucas Jones	Danielle Olson
Brent Barron	Kekuhi Kealiʻikanakaʻoleohaililani	ʻŌiwi Parker Jones
Scott Benesiinaabandan	Megan Kelleher	Caroline Running Wolf
Michelle Brown	Suzanne Kite	Michael Running Wolf
Melanie Cheung	Olin Lagon	Marlee Silva
Meredith Coleman	Jason Leigh	Skawennati
Ashley Cordes	Maroussia Levesque	Hēmi Whaanga
Joel Davison	Jason Edward Lewis	Tyson Yunkaporta
Kūpono Duncan	Keoni Mahelona	
Rebecca Finlay	Caleb Moses	

Day 2 Afternoon

The afternoon was spent in discussion of the five themes, and then reviewing the plan for further work after we departed.

Workshop 2

The second workshop was an eight day writing/artist residency. This was held in two private residences in the Kahala area of Honolulu. The focus here was on producing texts that responded to the concerns raised in Workshop 1 and subsequent conversations. See the participants' agenda in appendix 6.4 at the end of this document.

The agenda for this workshop was much looser, as it was conducted primarily as a residency, i.e., individuals and small groups working together to complete their contributions. We held a group review on day 1, and then had group check-ins at the end of most days so we could all inform each other of our projects. We split into three main groups:

- Prototype: this group developed a prototype of the Hua Kiʻi language app.
- Vignette: this group developed individual creative and speculative future imaginaries.
- Context: this group developed texts providing context for the entire project as well as organized the position paper as a whole.

By the end of Workshop 2, we had rough drafts of most components of what is now this position paper.

Workshop 2 Participants

Workshop 2 participants

Noelani Arista

Scott Benesiinaabandan

Michelle Brown

Melanie Cheung

Joel Davison

Suzanne Kite

Jason Edward Lewis

Caleb Moses

Issac ʻIkaʻaka Nāhuewai

Kari Noe

Caroline Running Wolf

Michael Running Wolf

Hēmi Whaanga

3.2

AI: A New (R)Evolution or the New Colonizer for Indigenous Peoples?

Dr. Hēmi Whaanga

"It's a familiar story these days: the era of Artificial Intelligence (AI) has arrived, and AI will soon render human labor and decision making obsolete." [1]

We are often told that there is one constant in life and that is change; it is inevitable, inescapable. When the forces of power begin to blow and conditions are right, change will happen. As the planet undergoes a period of transformation brought about by the advances of data science and the convergence of technologies, the Internet of All Things and AI, the potential of AI to be change agent for Indigenous peoples is a thought-provoking, and to a certain degree, daunting proposition. The rapid progress of technology and innovation, in terms of its volume, complexity, and exponential growth in computing

---

[1] Mateescu, A., & Elish, M. C. (2019). *AI in context: The labor of integrating new technologies* (Data & Society report), p 8. <datasociety.net/wpcontent/uploads/2019/01/DataandSociety_AIinContext.pdf>.

power, have drastically changed how we socialize, communicate, access, share, distribute and view knowledge and information. Is AI inevitable, inescapable, a fait accompli for Indigenous peoples?

Knowledge and information are the intellectual capital generated by families, communities, tribes and knowledge holders over multiple generations. This intellectual capital, our Indigenous knowledge systems, are a holistic, dynamic, innovative, and generative system that is embedded in lived experience. [2] Carried and embedded in stories, song, art, place names, dance, ceremonies, genealogies, memories, visions, prophesies, teachings and original instructions, these systems are passed orally from one generation to another. Unfortunately, Indigenous peoples, their languages and cultures are exceptionally vulnerable to the impacts of change, to globalisation, and its underlying goal to create a global village based on cultural, social, political and economic homogenization.

With homogenization comes loss. It has been suggested that by the end of this century at least 50% of the world's languages will face the prospect of death. [3] Many if not a majority of these languages will unfortunately be Indigenous languages. When we lose a language, we lose the conduit to our linguistic and cultural ecosystem. If we lose those ecosystems, we lose our identity, our history, our culture, and ultimately, we lose our power. [4] With the increase in the probability of this homogenization, will AI accelerate this change, this loss?

In a recent gathering of cultural and technological experts in Aotearoa I asked a number of broad questions to garner thoughts and reflections on Māori protocols, world views, technology and innovation. [5] Our discussions focused on the impact of new technologies, including virtual, augmented and mixed realities, AI and machine learning based on cultural language and knowledge. The discussion ranged across numerous fascinating topics such as data sovereignty, control, access, context, management, storage, and futures; IP and copyright; algorithms and attempts to decolonize them, Māori coding practices, Māori platforms, Māori AI and Māori life-force; safeguarding knowledge systems; and handling embedded biases and racism. Amongst the many responses, one statement from Professor Rangi Matamua stood out from the rest of the discussion: *Is AI the new (r)evolution or the new colonizer for Indigenous peoples?* This probing statement drew my attention and all of those in the room with me. We wondered: can an intelligence, or an artificial one at that, be used to colonize something or someone else?

---

[2]   Smith, L. T., Maxwell, T. K., Puke, H., & Temara, P. (2016). Indigenous knowledge, methodology and mayhem: What is the role of methodology in producing indigenous insights? A discussion from Mātauranga Māori. *Knowledge Cultures,* 4(3), pp. 131-156.

[3]   Thomason, S. G. (2015). *Endangered languages* (Vol. 1). Cambridge, UK: Cambridge University Press.

[4]   Nuwer, R. (2014, 6 June). *Languages: Why we must save dying languages. BBC.* <bbc.com/future/story/20140606-why-we-must-save-dying-languages>.

[5]   See <sftichallenge.govt.nz/research/atea>.

Colonization is often described as the act of invading and taking control by force, the act of taking something over for your own use, or the process of settling among and establishing control over the Indigenous people of an area. It subscribes to the language of appropriation, conquest, invasion, occupation and suppression. The colonization of the culture, language and mind takes place "through the transmission of mental habits and contents by means of social systems other than the colonial structure. For example, via the family, traditions, cultural practices, religion, science, language, fashion, ideology, political regimentation, the media, education, etc."[6] Theorists such as Frantz Fanon have written about the perpetuation of the colonial agenda in consciousness while Indigenous scholars like Linda Tuhiwai Smith and Ngũgĩ wa Thiongo have argued for the decolonization of our mental universe.[7] Thiongo wrote in *Decolonising the Mind*:

> the most important domination was the mental universe of the colonised, the control, through culture, of how people perceived themselves and their relationship to the world. Economic and political control can never be complete or effective without mental control. To control a people's culture is to control their tools of self-definition in relationship to others."[8]

Indigenous peoples across the world have faced and continue to face the effects of colonization, of mind control. As an example of modern day colonialism, consider Cambridge Analytica, the political analysis firm at the center of the Facebook data scandal. Cambridge Analytica harvested the personal data of millions of people's Facebook profiles without their permission and used that data for purposes of partisan political advertising.[9] These types of unscrupulous behaviours exacerbate existing societal biases, deepen inequalities, and contribute to the deterioration of trust across society. In response, a global dialogue has emerged that seeks solutions to these types of behaviours to enhance and improve economic, societal and environmental well-being. A range of documents and reports, aimed at setting global principles and standards governing AI, have been developed in order to establish corporate compliance, achieve industrial competitiveness, or to ensure sustainable development.[10] A scan through these numerous codes and declarations highlights a focus on common good and benefit for humanity;

---

[6] Dascal, M. (2009). Colonizing and decolonizing minds. In I. Kuçuradi (Ed.), *Papers of the 2007 World Philosophy Day* (pp. 308-332). Ankara, Turkey: Philosophical Society of Turkey, p. 309. <m.tau.ac.il/humanities/philos/dascal/papers/Colonizing and decolonizing minds.doc>.

[7] See Fanon, F. (1990). *The wretched of the earth*. London, UK: Penguin; Thiong'o, N. (1986). *Decolonising the mind*. Portsmouth, N.H.; Harare: Heinemann Educational; Zimbabwe Publishing House; and Smith, L. T. (2012). *Decolonizing methodologies*. London, UK: Zed Books.

[8] Thiong'o, N. (1986). *Decolonising the mind*. Portsmouth, N.H.; Harare: Heinemann Educational; Zimbabwe Publishing House, p. 16.

[9] Crabtree, J. (2018). *Cambridge Analytica is an 'example of what modern day colonialism looks like,' whistleblower says*. CNBC. <cnbc.com/2018/03/27/cambridge-analytica-an-example-of-modern-day-colonialism-whistleblower.html>.

[10] See Renda, A. (2019). *Artificial Intelligence – Ethics, governance and policy challenges (Report of CEPS Task Force)*. Brussels: Centre for European Policy Studies. <ceps.eu/wp-content/uploads/2019/02/AI_TFR.pdf>;

the establishment of principles of fairness and intelligibility; data and privacy rights; shared benefits and restrictions or outright bans on vesting AI with the autonomous power to hurt, destroy or deceive humans [11] However, Indigenous rights, issues and concerns are rarely discussed as part of this global dialogue apart from a recent report prepared by the Australian Council of Learned Academies (ACOLA), that discussed wellbeing, equity, self-determination and Indigenous data sovereignty. [12]

> We stand on the brink of a technological revolution that will fundamentally alter the way we live, work, and relate to one another. In its scale, scope, and complexity, the transformation will be unlike anything humankind has experienced before. We do not yet know just how it will unfold, but one thing is clear: the response to it must be integrated and comprehensive, involving all stakeholders of the global polity, from the public and private sectors to academia and civil society. [13]

To return to the main question of this essay: 'Is AI the new (r)evolution or the new colonizer for Indigenous peoples?' This type of question and the impacts of colonization and moral and ethical boundaries is not something that can be answered in the context of a single paper or a single conversation. It will, however, be something that this generation, often referred to as 'digital natives,' 'homo zappiëns,' 'Net generation,' 'millennials,' 'i-generation'—a generation raised, immersed and exposed to a myriad of digital technologies— will have to tackle. AI will be a game changer that challenges the foundations of our knowledge systems. Thus, it is critically important that we envision and shape how AI could be part of a revolution that is productive for our knowledge systems, our languages, and our futures. We need to be part of the dialogue on establishing global principles and standards for the use of AI to ensure that is not used to perpetuate societal biases, inequalities and global homogenization.

References

Crabtree, J. (2018). *Cambridge Analytica is an 'example of what modern day colonialism looks like,' whistleblower says. CNBC.* Retrieved from cnbc.com/2018/03/27/cambridge-analytica-an-example-of-modern-day-colonialism-whistleblower.html

Dascal, M. (2009). Colonizing and decolonizing minds. In I. Kuçuradi (Ed.), *Papers of the 2007 World Philosophy Day* (pp. 308-332). Ankara, Turkey: Philosophical Society of Turkey. Retrieved from m.tau.

---

[11] Walsh, T., Levy, N., Bell, G., Elliott, A., Maclaurin, J., Mareels, I.M.Y., Wood, F.M. (2019). *The effective and ethical development of artificial intelligence: An opportunity to improve our wellbeing* (Report for the Australian Council of Learned Academies, acola.org). Melbourne, Australia: Australian Council of Learned Academies. <acola.org/wp-content/uploads/2019/07/hs4_artificial-intelligence-report.pdf>.

[12] See Renda, A. (2019). *Artificial Intelligence.*

[13] Schwaub. K. (2016). *The fourth industrial revolution: what it means, how to respond. World Economic Forum* [para. 1]. <weforum.org/agenda/2016/01/the-fourth-industrial-revolution-what-it-means-and-how-to-respond>.

ac.il/humanities/philos/dascal/papers/Colonizing and decolonizing minds.doc.

Fanon, F. (1990). *The wretched of the earth.* London, UK: Penguin.

Mateescu, A., & Elish, M. C. (2019). *AI in context: The labor of integrating new technologies* (Data & Society report). Retrieved from datasociety.net/wpcontent/uploads/2019/01/DataandSociety_ AIinContext.pdf.

Nuwer, R. (2014, 6 June). *Languages: Why we must save dying languages. BBC.* Retrieved from bbc.com/future/story/20140606-why-we-must-save-dying-languages.

Renda, A. (2019). *Artificial Intelligence – Ethics, governance and policy challenges (Report of CEPS Task Force).* Brussels, Belgium: Centre for European Policy Studies. Retrieved from ceps.eu/wp-content/ uploads/2019/02/AI_TFR.pdf.

Schwaub. K. (2016). *The fourth industrial revolution: what it means, how to respond. World Economic Forum* [para. 1]. Retrieved from weforum.org/agenda/2016/01/the-fourth-industrial-revolution-what-it-means-and-how-to-respond.

Smith, L. T., Maxwell, T. K., Puke, H., & Temara, P. (2016). Indigenous knowledge, methodology and mayhem: What is the role of methodology in producing indigenous insights? A discussion from Mātauranga Māori. *Knowledge Cultures,* 4(3), 131-156.

Smith, L. T. (2012). *Decolonizing methodologies.* London, UK: Zed Books.

Thiong'o, N. (1986). *Decolonising the mind.* Portsmouth, N.H.; Harare: Heinemann Educational; Zimbabwe Publishing House.

Thomason, S. G. (2015). *Endangered languages* (Vol. 1). Cambridge, UK: Cambridge University Press.

Walsh, T., Levy, N., Bell, G., Elliott, A., Maclaurin, J., Mareels, I.M.Y., Wood, F.M. (2019). *The effective and ethical development of artificial intelligence: An opportunity to improve our wellbeing* (Report for the Australian Council of Learned Academies, acola.org). Melbourne, Australia: Australian Council of Learned Academies. Retrieved from acola.org/wp-content/uploads/2019/07/ hs4_artificial-intelligence-report.pdf.

3.3

The IP AI Workshops as Future Imaginary

Jason Edward Lewis

> The future is happening
> It just hasn't reached us
> Yet.
>
> —Scott Benesiinaabandan[1]

The Indigenous Protocol and Artificial Intelligence Workshops (IP AI) are a way of "practicing the future together."[2] By inhabiting physical, emotional and intellectual space, IP AI provides a much-needed context in which we can take wisps of whimsy and filaments of fancy and weave them together with the rough cords of our contemporary struggles and the thick braids of our ancestors' dreams to make new realities material. Such spaces are few, and those few are rarely found at the intersection of

---

[1] Benesiinaabandan, S. (22 May, 2019). Personal communication.

[2] brown, a.m. (2017). *Emergent Strategy: Shaping Change, Changing Worlds.* Chico, California: AK Press, p. 32.

Indigenous life and the world of Western-dominated technological transformation.

"What we pay attention to grows," writes adrienne marie brown (2017). The question is, "how [do] we grow what we are all imagining and creating into something large enough and solid enough that it becomes a tipping point"?[3] Our aim with these workshops is to create something large enough and solid enough that Indigenous people become central participants in shaping the future of artificial intelligence systems, and— by extension—the future of our technology-saturated world. We aim to build a set of future imaginaries where our everyday interactions with technology are characterized by a compatibility—a deep integration— between our cultural protocols and the protocols determining how that technology operates. Our aim is to foster a productive resistance, a refusal to accept that all that is solid melts into air. And to go further, to find firm footing in Indigenous cultures tested by a half millennia of colonialism and use them to launch ourselves (yet again) into the future.

The IP AI conversations have been expansive and deep. Expansive, in that they cover extensive ground that includes epistemology, culture, machine learning, colonization, temporal models, ontology, software architecture and linguistics. Deep, in that they dig down through layers of Indigenous history, language and culture from the position of particular Indigenous individuals and their communities. We use the term 'Indigenous' as connective tissue rather than descriptive skin, to appreciate the hyperdense textures of our points of contact while respecting our rich and productive differences.

This is how the future begins: by thinking anew.

Over the course of our workshops, our conversations took place in Indigenous-determined spaces, with mostly Indigenous participants, and in a territory where Indigeneity is present at every turn. Neuroscientists traded ideas with cultural knowledge-holders, who traded ideas with computer scientists, who traded ideas with poets, who traded ideas with language-keepers, who traded ideas with visual artists, who traded ideas with hula teachers, who traded ideas with historians, who traded ideas with engineers. The challenge was real: it is all too easy to concede incommensurability in the face of such a variety of disciplines, cultures and politics. But we talked and we ate and we shared stories about ourselves, our peoples and our practices to build scaffolding between us.

This is how the future comes into view.

We dreamed about tomorrow, and the day after, and 500 years later. We observed protocol together; we ate together; we chanted and sang together. We mapped paths forward that draw on our peoples' long histories of technical innovation and scientific practice, sharing examples of how our traditions offer a wellspring of inspiration for engaging with the world and with each other through the tools we make.

---

[3]   ibid.

This is how the future gets sketched out.

We spent first two days and then, three months later, ten days dwelling in a future-present-past, expressing sovereignty using 360-degree seeing that ranged across disciplines. Anishinaabe participants talked about how *oskabewis*, helpers whose generous and engaged and not:invisible support for those participating in ceremony, could model how we might want AI systems to support us—and the obligations that we, in turn, would owe them. Hawaiian participants talked about all the steps involved with crafting a fishing net, the layer upon layer of permission and appreciation and reciprocity required to properly work with those relations—expressed through prayer, chant, and song—protocols that could model how we might want to create our hardware and software systems from a foundation of ethical care. Maori participants talked about concerns in their communities about how knowledge will get passed down to the children and grandchildren, and speculated with us about holographic aunties who would work with members of the community to preserve and transmit that knowledge. Coquille participants talked about embedding their cultural values of care and trust into AI systems integrated with blockchain technology to help the tribe make decisions about sharing and then distributing community resources. We discussed Blackfoot metaphysics, and the implication from Leroy Little Bear's writings that Blackfoot might be the best language in which to work on quantum physics, and imagined what other isomorphies might exist between specific Indigenous languages and scientific frameworks, and how recognizing and leveraging such resonances might provide insight into the great technical challenges of our time. [4]

This is how the future gets filled in.

We considered different layers of the stack: hardware architectures and software protocols that make high-level computation possible, and how, as we move first up the hardware stack from silicon to circuits to microchips to computers to networks; and then up the software stack from machine code to programming languages to protocols to systems, how each of those layers is culturally inflected. We wondered what would happen if that culture was an Indigenous one—microchips produced with the care of a Lakota community raising a sweat lodge; computers constructed with the intentionality of a Cree singer building his hand drum; networks knitted together following Coquille practices for making woven cattail trays; a programming language written in Crow to reflect Crow understandings of data and process; an operating system designed by Cheyenne computer scientists; pattern recognition algorithms taught using Aboriginal techniques for creating songlines; governance expert systems following Haudensonee political formations; an AI nurtured on Kanaka Maoli concepts of ʻāina, ʻohana and kuleana.

This is how the future gets prototyped.

We asked our questions—not the questions of the colonizer. How will these devices be made? Who will

---

[4]   Little Bear, L., and Head, R.H. "A Conceptual Anatomy of the Blackfoot World." ReVision, vol. 26, no. 3, Winter 2004, pp. 31–38.

make them? With whom will they be in relation once they are in the world? How will they conduct themselves as relations in our communities? How will our communities treat them as relations? To what ends will they be shaped? How will they help our communities grow and thrive? How will our non-human kin take to them? Will they be there for our seventh-generation descendants? For many of us working in or with experience of the high-tech industry, it was a relief to focus on such questions rather than the tired tropes of a technology elite that recursively chases its own tail upon a ground of epistemological blindness, cultural prejudice and myopic misanthropy. Asking our questions allowed us to thread together what we know within our communities with what we are still learning. Asking our questions shows our youth how our knowledge frameworks can provide the tools to inquire incisively about the world to learn from it, and how to better live in it. Asking our questions asserts our sovereignty, over our minds, our lives and our futures.

This is how the future reaches us.

References

brown, a.m. (2017) *Emergent Strategy: Shaping Change, Changing Worlds*. Chico, California: AK Press, 2017.

Little Bear, L., and Head, R.H. "A Conceptual Anatomy of the Blackfoot World." *ReVision*, vol. 26, no. 3, Winter 2004, pp. 31–38.

SECTION 4

Vignettes

Kamapuaʻa/kalo. Image by Kūpono Duncan, 2019.

4.1

Gwiizens, the Old Lady and the Octopus Bag Device

Scott Benesiinaabandan

adizookaan - sacred stories - ontologies
agwanem - hold in the mouth - structure/shapes
mamawi - to be/work together - intra/interactions
booshke giin - it's up to you - sovereignty

"If it is a human thing to do to put something you want, because it's useful, edible, or beautiful, into a bag, or a basket, or a bit of rolled bark or leaf, or a net woven of your own hair, or what have you, and then take it home with you, home being another, larger kind of pouch or bag, a container, you take it out and share it or store it up for winter in a solider container or put it in the medicine bundle or the shrine or the museum, the holy place, the area that contains what is sacred, and then the next day you probably do much the same again-if to do that is human, if that's what it takes, then I am a human being after all. Fully, freely, gladly, for the first time." [1]
—Ursula Le Guin

---

[1]   Le Guin, U. K. (1989). The Carrier Bag Theory of Fiction. In *Dancing at the Edge of the World: Thoughts on Words, Women, Places* (pp. 165-170). New York, NY: Grove Press. pp. 151-152.

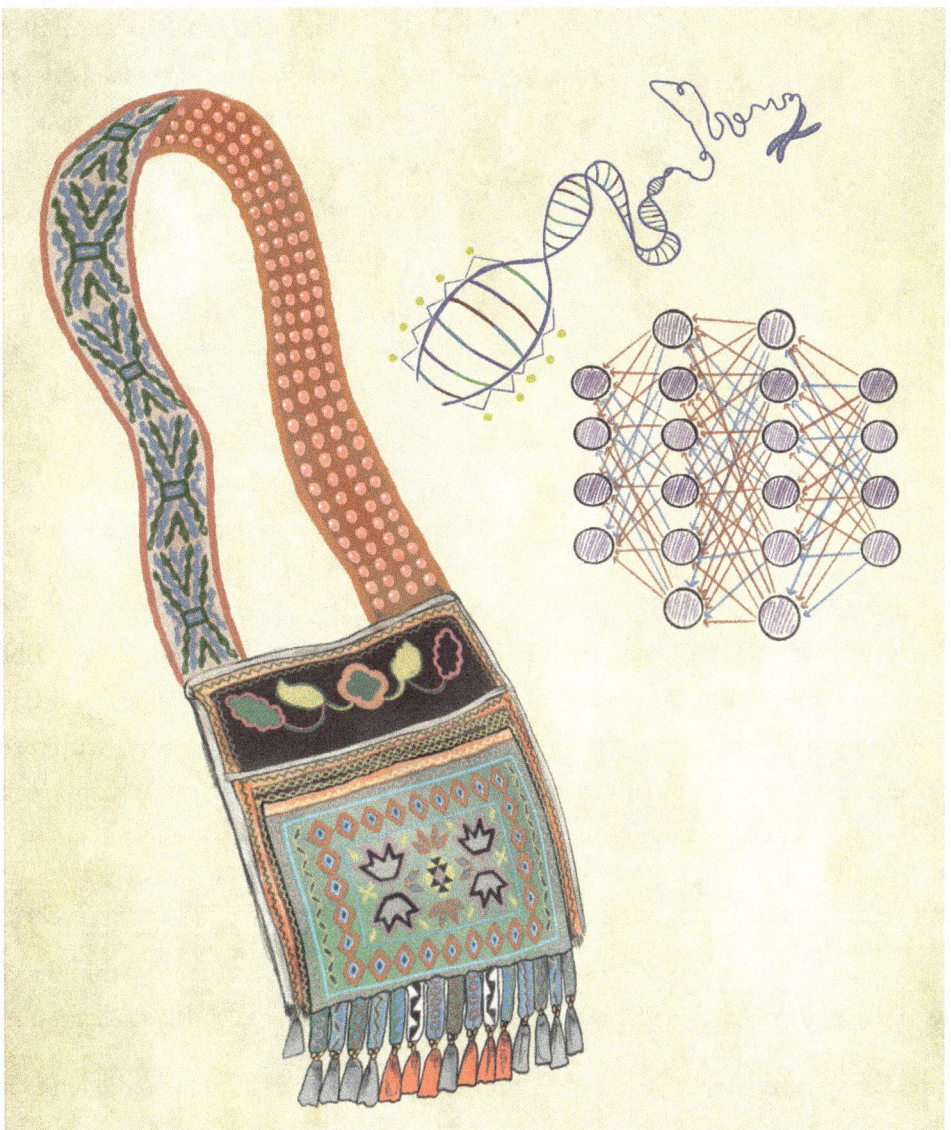

Octopus Bag. Image by Kari Noe, 2019.

The Octopus Bag Device is a removable and relatively non-invasive (it can be removed without issue), multisensorial computing device that uses our inherent DNA capabilities, both as storage and computational power (using true parallel computing). This device is held in the mouth by the molars along the side of teeth and against the cheek—both teeth and cheeks are fertile areas to access DNA material. Elements of the device are in direct contact with the back of the jaw where our 'wisdom' teeth used to (accounting for our skeletal evolution) come in. This allows the device to have direct skeletal vibration conduction. Other elements extend up from the mouth and sit loosely in the nostrils to allow for pheromonal stimulations. Sensory inputs (smell, sight, taste, touch, hearing) are stored in universally unique and overlapping ratios to the individual.

Far-future DNA (true parallel) computing and mass storage removes the issues of land-based resources. Each person's DNA-landscape is at some level unique and therefore the input and output of the computational + storage systems would also be unique and offer far greater power, security and sovereignty than one-computer+OS fits all solutions of today. The device is also an interface using our personalized interior modes of perception to interpret and evaluate the computational outputs. Specific AI programs are developed through the new possibilities presented by the nature and power of this type of computing.

The Octopus or Bandolier bags are a long standing tradition across Turtle Island, most notably within Algonquin speaking communities. They originally were used for holding medicines and/or fire making utensils. In later years (post 1700s), they became increasingly decorated and valued, with people often carrying two overlapping bags. Originally, the bags were inverted animal skins including legs and tails, which led to the subsequent stylization of the longer tabbed bags, thus having the octopus appearance. Four sets of two tabs fasten together at the ends with cloth or bells (think regalia bells...the sound of which drives of 'bad' spirits), often decorated on both sides. In this story, the shape of the Octopus bag is a multi-level metaphor for the overarching story structure. Some elements found within the Octopus bag device, and its associated AI, 'shapes' the way knowledge permeates throughout our communities.

In this story, the Octopus bag device is found, wrapped in red cloth, inside a blue and white quilled octopus bag.

Ningoding ayindaawag...
[once upon a time...]

I remember in my mid-twenties, finishing my last year of my undergraduate degree, sitting in the University of Winnipeg's rare book room and reading a small story. This small story I accidently found was buried in a book I can no longer reference, recall or actually find though I have tried in earnest more than a few times. And although I cannot find that book, what is left with me is a vivid memory of reading this very small story, and the big impact it has on me.

Two decades have passed since then and I still have this experience stuck in my head. The story itself, the feel of the book, a few illustrations, the light of the day, the table I sat at. At that particular moment in time [1 April 2001] nothing about this story had much to do with the practical application of AI (both current and future-speculative). But that small story had left a deep impression on me regarding our complex relationship with technologies. That the story could serve as a map to help us navigate our relationship with AI wasn't fully clear to me until taking part in the Indigenous Protocol and AI Workshops [2019].

Since 2001, I've frequently thought of this very small story in big ways. I have incorporated a diversity of technologies, common and experimental, in both my day-to-day life and artistic practice. Over the years, I have read and listened to hundreds of *adizkookaan* [sacred stories] told in this same way, with a similar cast of characters. What has resonated most was the person/being in the forest with an immense gift to give after a conflict or contest. The inter-relationship to tech (broadly) and AI (specific to this project) and my art practice has always been rooted in this story, both overtly and subtly, consciously and subconsciously.

When presented with the opportunity to write/make something about AI, this small story was there waiting. I chose not to go back to find the original story, but to build a story based on the materiality of memory and my long relationship with other, similar stories I have read and heard.

 The following story is a story rooted in a forgotten archival source, itself ripped from future imaginaries articulated in deep historical time, and then recalled in my own way for me to draw upon as an artist and an Anishinaabe in the nowtime of this project.

Ningoding ayindaawag, miinawaa...
[once upon a time, once again....]

...there was a village and in this village there was a Little Boy. No one really remembers how the Little Boy came to the village, not even the old man and old woman who the Little Boy sometimes called *Ninga and Noos* and sometimes *Kookum and Mishoomis*. His arrival to this village could not have been too long in the past, for he is yet still a Little Boy. But the *adizkookaan* and the *memoryspacetime* where they reside are strange things, and people in the village that the Little Boy now called home never asked these sorts of questions.

It came to pass that the Little Boy's village slowly became stricken with an unknown killing-sickness. Despite all the best attempts by the medicine people to find a cure, it was for naught and soon the Little Boy was the only one left unaffected.

Now, the Little Boy loved all the people in this village and it hurt his heart to see his village so sick. After some days of deep contemplation, the Little Boy knew he had to try his best to find someone that could help his people. The next morning, gathering up his resolve, the Little Boy set out at dawn to seek help.

Most of the day, the Little Boy made his way along the well worn path that he had been down many,

many times before. This time however, the forest was strangely silent, and he heard and saw no one that might help his village.

Near the end of the day, the Little Boy was getting very tired and he began looking for a place to rest. Just then he heard the sound of someone singing.

The Little Boy can't quite recognize the sound of the voice.
The Little Boy cannot quite recognize the language of the song.

Compelled and curious, he investigated the source of the song.

The Little Boy finds an Old lady sitting alone and tossing stones into a fast flowing river. The Little Boy quietly watches:

pick toss arch

( )

pick toss arch

(( ))

pick toss arch

((( )))

pick toss arch

(((( ))))

After watching her for a while, the Little Boy enters the clearing and as he gets closer Little Boy can see that the Old Lady is both very old and impossibly straight and strong. He also sees that she is carrying a brilliantly coloured blue and white quilled octopus bag that hangs over her sharp shoulders and down her thin sides. It is adorned with a design that he doesn't recognize, and at the bottom of this bag eight tabs hang down with tiny copper bells that reverberate strangely in his ears.

The Old Lady also wears a grin, her mouth is full of sharp and shining teeth.

The Old Lady notices the Little Boy looking at the blue-white quill bag.

The Old Lady asks what brings such a Little Boy along this path, for this is a long distance from his village.

The Little Boy tells the Old Lady about his village, about the sickness that is killing all of his people and his journey to find help to save his village.

_____ |*a whiskey jack is cackling near the edge of the clearing*

The Old Lady nods. She takes off the beautiful quilled bag and lays it on the ground. From it, the

Old Lady takes out a small object, carefully wrapped in red cloth. The Old Lady carefully unwraps the object.

The Little Boy looks at the object, now sitting exactly in the middle of the red cloth. It glints in the sun, like the Old Lady's sharp teeth. The Little Boy can't recall seeing anything quite like it before.

> *Pheromonal memory and acoustic visualizations based on iconic(visual) and echoic(acoustic) memory storage and retrieval for the interior interface. The device uses endogenic enhancements for the interface. Small doses of original tobacco/asemaa ratios can be injected to strengthen the interior multi-sensorial interface.*

The Old Lady says that this shining-object-on-the-cloth will certainly cure his people and she is very willing to help the Little Boy's village, *"But"* the Old Lady says, pointing at the Little Boy with her lips, *"there is one thing you must do first"*.

The Old Lady says that the Little Boy must first find, wrestle and defeat her to gain this prize. The Old Lady says he will have three tries and if by the third try the Little Boy cannot pin her to the ground, the Old Lady will devour the Little Boy.

The Little Boy moves back a step—the Old Lady's sharp, shining teeth becoming suddenly a lot more dangerous. Then after considering it, the Little Boy agrees to the challenge, after all, *"How hard can an Old Lady be to defeat"*, he thinks to himself, *"This is his forest and afterall, a Little Boy with great strength and speed and she is just an Old Lady"*.

He is instructed to prepare himself for four days, after which time he must set out to find and begin his contest with the Old Lady with the mouthful of impossibly sharp and shining teeth.

"Continue looking for me until you hear the sound of a whiskeyjack, smell the scent of strawberries, hear the crack of distant thunder and see where the poplar trees turn their silver leaves upwards and inside out... soon after you will find me and we will begin this contest".

The Old Lady gently touches the beautiful blue and white bag with intricate designs and turns away from the Little Boy. The Old Lady resumes singing

an Almost Recognizable Song
 in
   an Almost Recognizable Language

The Little Boy turns toward the sunset to leave but realizes he is now quite lost and it is twilight by the time Little Boy finds the return path home. Walking through the night, the Little Boy finally reaches home just as dawn breaks across his village.

The Little Boy's wigwam is on the farthest side of the dark and still village. As he makes his way, his ears fill themselves with the sound of moans and soft sobs of his sick relatives.

Exhausted, the Little Boy finally reaches his wigwam and quickly falls asleep.

Over the next four days, the Little Boy doesn't tell his people about the Old Lady in the forest, nor of his challenge to win the device that will cure everyone of their sickness.

For the next four days, the Little Boy eats very little, taking only a sip of tea at each sunset. As the day comes to depart, the Little Boy gathers up his finest moccasins (made for him by his mother), his strongest arrow heads (made for him by his grandmother), and takes his finest knife (given to him by his grandfather) and, putting it all into a bag, sets out to locate the Old Lady.

Striking out at dawn, the Little Boy begins his search for the Old Lady.

No one really knows how long the Little Boy searched, maybe a day, maybe a month, maybe longer but certainly the Little Boy travelled very far and was very tired by the time he smelled a

faint scent of strawberries

Recalling the Old Lady's instructions,
he stopped and immediately heard the

cackling laugh of a whiskey jack

coupled with the

distant sound of animiikiikaa

The Little Boy, turning in a circle, realized
that all of the poplar trees had turned their

green and silver leaves
upside down and inside out

The Little Boy knew he was now close.

Continuing down the path a little way, the Little Boy heard the same almost familiar song sung in the almost familiar language of the Old Lady. At the same time, he heard a booming voice behind him in a tree, *"Where are you going, Little Boy?"*.

Turning around, he traced the very big voice to a very small tree frog. The Little Boy tells the Frog his story. Frog says he knows of this Old Lady and warns the Little Boy that he likely can't defeat this

Old Lady, and certainly not without Frog's help. Frog proceeds to tell the Little Boy of a plant he can use to help him in his fight. Reaching into his impossibly small frog bag, he lifts out a little ball of medicine and gives it to the Little Boy with the instruction,

"When you think you are close to defeated, eat a little of this and it will certainly help you".

The Little Boy takes this medicine from Frog and puts it carefully into his bag.
"Miigwech Chi' Omagagii!" says the Little Boy as he proceeds down the trail.
"Baamaapii, Little Boy!", Frog calls out after him.

Entering the clearing, the Old Lady is there standing, waiting and ready with her beautiful blue and white bag wrapped around her thin shoulders and her mouthful of teeth—impossibly sharp and shiny.

The two begin to grapple, the Little Boy immediately realizes this Old Lady is far stronger and more clever than she looks, *"Maybe even stronger and trickier than myself,"* thinks the Little Boy.

grapple wrestle trip fight trip wrestle grapple
grapple wrestle trip fight trip wrestle grapple
grapple wrestle trip fight trip wrestle grapple
grapple wrestle trip fight trip wrestle grapple

For a long time they wrestled, no one remembers exactly how long.

After trying all of his tricks of speed and strength, the Little Boy realized that he cannot win this match. Desperately reaching into his bag, he grabs some of the medicine Frog gave him, puts it in his mouth and awaited victory.

The plant doesn't appear to do anything to bolster the Little Boy's losing fight, no strength comes to his worn out arms and no speed to his worn out legs.

The Old Lady, looking down with her sharp toothed impossible grin, easily pins the Little Boy to the ground and wins the contest.

"You see how foolish you are Little Boy, certainly you see that I will win and devour you. Let me just get it over with now and you can save yourself the sadness of seeing your people die."

"Gawiin" says the little Boy as he dusts himself off. *"I will certainly defeat you next time."*

Old Lady shrugs a nonchalant shrug and says to return in another four days and try again, while instructing him to *"Keep looking for me until you hear the sound of a whiskeyjack, smell of scent of strawberries, hear the crack of distant thunder and see where the green and silver poplar trees turn their leaves upwards in inside out, soon after you will find me and we will fight once again."*

Data sharing and Interface with other humans + non-humans is achieved with a left arm based interface, based on the positioning of the heart on the left side of the body and the original teaching of how to interface (collect, share and offer) with tobacco.

This time, the Little Boy returns to his community at twilight. He makes his way through the village to his wigwam. The thickening sounds of sickness are heard throughout.

This time his grandfather is awake. *"Where have you been Giiwenz?"*

Not able to hide it any longer, the Little Boy tells his grandfather about his intolerable sadness, about his people dying, his search for the cure, the Frog, the Old Lady, the Octopus Bag Device that will save the village. He shows his grandfather the medicine the Frog had given him. *"Give me a little piece of that, Giiwenz, and let me sleep."*

For the next four days the Little Boy prepares for his challenge. This time he knows how strong and tricky Old Lady is and plans accordingly. All day, the Little Boy prepares for his second challenge, and every night the Little Boy sits with his grandfather at their fire, listening to him recount his own stories of when he too was just a Little Boy.

At dawn on the fourth day, the Little Boy awoke and headed out of the village to begin another search for the Old Lady.

No one really knows how long the Little Boy searched, maybe a day, maybe longer but certainly the Little Boy was very tired by the time he smelled a

f a i n t s c e n t o f s t r a w b e r r i e s

Recalling the Old ladies instructions, he stopped and could immediately hear the

c a c k l i n g l a u g h o f a w h i s k e y j a c k

and the

d i s t a n t s o u n d o f a n i m i i k i i k a a

The Little Boy turning in a circle realized that all of the silver poplar trees had turned their

g r e e n a n d s i l v e r l e a v e s

u p s i d e d o w n a n d i n s i d e o u t

He knew he was close once again.

Continuing down the path a little way, the Little Boy heard the same *almost familiar song sung in the almost familiar language* of the Old Lady. At the same time the Little Boy heard a familiar voice behind him in a tree.

"Boozhoo boozhoo, Giiwez, Aniin, Little Boy! I see you did not defeat the Old Lady last time? I see, I see. You probably took the medicine at the wrong time and therefore it wasn't strong enough to help. No matter, Giiwenz, I have another plan to help you."

With that Frog calls a deep booming frog call. **<>echoes<>**

<>echoes<>

<>echoes<>

In a very short time, the Little Boy hears someone running down the path. A black wolf appears.

"Boozhoo Frog" says Maa'ingaan, *"Boozhoo Wolf'"* says Omaagaakii. *"Boozhoo Little Boy"* says Wolf. *"Boozhoo, Boozhoo"* says Giiwenz.

Frog tells Wolf about the Old Lady, the beautiful blue and white bag, the device in red cloth, the sickness of his village and the impossible challenge to defeat the Old Lady. *"Can you help this Little Boy, Maa'iigan?"* asks Frog in his big voice.

"Ah, hahaha...eya, no worries," says Wolf, *"For I am silent, and smell of the forest, I can move undetected by the Old Lady. I will scout ahead and let you know when her back is turned. When it turns away, you can then run in, for you are a young and a very, very quick boy and she is a very, very old lady, her hearing cannot be very good. You can then easily sneak up behind her, pin her to the ground and win the cure for your people."*

Quickly assessing Wolf's idea as quite clever, Little Boy readily agrees to it.

"Chi' miigwech Ma'iingan!" Little Boy called out as Wolf silently darted off, down the path towards the almost familiar song sung in the almost familiar language of the Old Lady.

> *New information is accessed alongside our ongoing and deepening DNA archive of memories, ontologies/knowledge bases [stories, songs, etc.], stretching as far back as our DNA remembers.*

No one remembers how long Wolf waited for the Old Lady to turn her back, it was certainly more than a day, and probably much longer until Wolf finally, sitting undetected in the deepest and darkest part of the forest shadows, saw the Old Lady turn her back. He quickly ran back to tell the Little Boy.

Little Boy was ready for Wolf's return and immediately struck off down the trail, sneaking towards the clearing which was, by the time he reached it, brightly illuminated by a full moon. The Little Boy could clearly hear Old Lady singing and clearly see the Old Lady with her back to him.

Little Boy runs as silently and quickly as possible toward the Old Lady but just as he reaches her, the Old Lady moves out of the way of Little Boy. Off-balanced, the Little Boy loses his footing, trips and lands in a pile of wolf shit into which the Old Lady easily pins the Little Boy to the ground and wins the second match.

The Little Boy sheepishly gets to his feet, spitting out the shit from his mouth. Laughing, the Old Lady tells the Little Boy that she has eyes that see in all directions, that no one in the world can sneak up on her, and most certainly not a Little Boy.

Knowing he only has only one more chance to save his village the Little Boy leaves the Old Lady, who has turned her back on him and had again continued her singing. After a while he finds a way back to the path he came from.

By the time he finds his way home his whole community has now fallen completely and terribly sick. Tonight there is no smoke rising from any of the homes as the fires have all burned out. Even the soft, fevered murmuring of sickness have all but ceased with only a cold and damp silence now wrapping his village.

Seeing this, the Little Boy knows he cannot wait four days. Taking one more look across his village, he turns back to the forest and seeks another path to find the Old Lady.

Again, the Little Boy searches for the signs of the Old Lady, but weakened by hunger, doubt and sadness, he knows he cannot go further. The Little Boy climbs a very tall and straight tree and falls asleep in the branches.

As he closes his eyes, he hears a whiskeyjack cackling from a nearby tree.

As the dawn breaks, the Little Boy wakes up and climbs down from his night-perch. His sleep was without dreams, no new ideas came to him as to how defeat the Old Lady. As he steps down from the tree, he hears the Old Lady singing just over the crest of a small hill.

The Little Boy approaches and sees her sitting on the edge of a lake.

The Old Lady slowly stands up turning as she speaks

"Little Boy, you are very early for our last and final contest. Be that as it may, you and I are here now and will fight. You can try one last time to save your people.

Remember, if you can pin me to the ground you get this—she taps the blue and white bag, with intricate designs—and if you lose, then I will eat every last piece of you and your people will surely die."

Left with no other options, the Little Boy decides to drop his belongings for they now feel much too heavy and he has no energy left to use them for this final contest anyway. The Little Boy casts down his favourite knife, the one given to him by his grandfather (*he can now barely recall his face*), he tosses down his sharpest arrows, the ones the ones his grandmother (*he can now barely recall her voice*) made for him, he takes off his favourite moccasins, the ones his mother (*he can now barely recall her smell*) made for him.

With the very last of his energy the Little Boy advances on the grinning Old lady, her mouthful of teeth shining like a sun—somehow brighter than each of the times before, *the Little Boy struggles to recall the first time he laid eyes on this Old Lady.*

Just as he reaches the Old Lady, the Little Boy realizes he is falling, time slows, the world is blinking in and out of phase, the edges of the Little Boy's vision is starbursting and rapidly oscillating towards a warm inviting black. All sounds are stretched out and hollow. The last thing he sees is the Old Lady with that strange beautiful bag standing over him grinning with her sharp teeth ready to devour the Little Boy's world.

Little Boy is awakened by a laughing whiskeyjack dropping rocks from a tree above him, hitting him directly and painfully on his head.

After a few moments he sits up and looks around. He realizes that he is still alive, that the Old Lady with the World-eating teeth is gone. But the beautiful blue and white bag with its intricate designs, its eight long tabs with those impossibly beautiful bells at the ends, the one that holds the device is now draped around his own shoulders and down past his own waist. With the greatest of hopes, he peers inside and sees the promised device carefully wrapped in red cloth, the one that will help save his people.

From just beyond the edge of the clearing the whiskeyjack still laughs and sings. It is the same song that the Old Lady was singing, this time however it is clear:

> *ogii-shawenimaan giche' manito* -- [*be kind to us great mystery*]
> *ogii-shawenimaan giche' manito* -- [*be kind to us great mystery*]
> *ogii-shawenimaan giche' manito* -- [*be kind to us great mystery*]
> *ogii-shawenimaan giche' manito* -- [*be kind to us great mystery*]

References

Anderson, M. (2017). *A Bag Worth a Pony. The Art of the Ojibwe Bandolier Bag.* Saint Paul, Minnesota: Minnesota Historical Society Press.

Le Guin, U. K. (1989). The Carrier Bag Theory of Fiction. In *Dancing at the Edge of the World: Thoughts on Words, Women, Places* (pp. 165-170). New York, NY: Grove Press.

Pomedli, M. (2014). *Living with Animals: Ojibwe Spirit Powers*. Toronto: University of Toronto Press.

4.2

Gifts of Dentalium and Fire:
Entwining Trust and Care with AI

Ashley Cordes

The world was a cold, dark, unnavigable place that needed warmth for survival and incandescence to see through the thick blackness that enveloped the land and air. To mutually resolve the problem, non-human animals of all forms helped by carrying in fire, small bit by small bit, leaving their noses and hooves forever blackened from the soot. Yet with the benefits of the technology of fire came an unexpected whoosh. It spread diseases, and the diffusion of carbon dioxide became akin to the spread of both viral and bacterial matter.

Each dentalium, units of tusk-like shells from the shores of the Pacific Northwest, are filled with computational fluid dynamics simulations. These show a high velocity jet of fluid being injected into a medium at rest. Each strand is dependent upon the genesis shell and its generational adaption. The black beads anchor the dentalium nodes within a distributed register maintained by the entirety of the network (necklace). This offering to AIs is intended to enable the externalization of stories/data/dreams which flow through the fluid in each shell. When used and worn, by AIs or otherwise, it is a symbolic means of sharing as well as an expression of regard for self and for others. Image by Kari Noe and Ashley Cordes, 2019

This legend passed on by some elders of Pacific Northwest Coastal Nations illustrates one version of how the technology of fire came to be. It provides insight into the need for Indigenous people to guard against the bad that comes alongside the good of certain technologies in our communities, and the need to be open to the help of spirited non-human beings. This can be accomplished through preparedness and involvement in the technological creation and decision-making needed for survivance [1] within the conditions of an information-saturated technoscape.

Two emerging technologies, AI and blockchain, are now being hyped as transformative agents in informational and medical industries as well as in the world of currency and record-keeping. However,

---

[1] Vizenor (1994) uses the term survivance to describe contemporary displays that show pride and tradition in the face of colonialism. See Vizenor, G. R. (1994). *Manifest manners: Postindian warriors of survivance*. Middletown, CT: Wesleyan University Press.

the potential around the coupling of AI and blockchain technology has not been adequately developed from Indigenous perspectives. This essay explores the potential of AI and blockchain to contribute directly to distribution, decision-making, and record-keeping for Indigenous communities' benefit, while weaving trust and care into the core of the conversation.

Blockchain and AI

'Blockchain technology' is most frequently described as a system of digital peer-to-peer assets enabled by software, secured using cryptography, and dependent upon a decentralized network for verification and distribution (Nakamoto, 2008). Peer-to-peer (P2P) processes allow multiple parties to transact without intermediaries such as governments or banks. 'Blockchain' specifically refers to an electronic record-keeping[2] (ledger) system that stores data and, in the case of cryptocurrency, records transactions using timestamps and hashes. Every time a transaction is made, financial or otherwise, a block of information is added to existing blocks of information to compose a chain that cannot be easily tampered with.

The ideas driving blockchain technology are generative in their capacity for expanding notions about how to decentralize control and increase trust in a system. Politically, there is less concentration of authority, and systematically, any threats to the security and privacy such as hacks or the selling of personal data can be minimized. Additionally, AI, technologies that exhibit the complexities of human and non-human intelligences can be paired with blockchain in productive and innovative ways. The coupling of technologies in many cases occur in a two-step process; the first involving AI in making informed and complex decisions, and the second with blockchain in recording the outcome of those decisions in a fashion that is ostensibly 'immutable.' AI, for example, can allow for processing and decision making using the data stored in blockchain. Symbiotically, blockchain can then provide a reliable record of the decisions that AI subsequently makes, allowing for the genealogy of decisions to be traceable.

Moving Toward an Indigenous Protocol, Data

AI is largely framed in consumer industries as tools or products to make life's wide range of tasks easier, quicker, and often what is perceived as better. Since AI is trained with data to do things such as reason, predict, and represent, data become the archives of profound significance and vulnerability. While easier access to intellectual/traditional/cultural data affords the opportunity for Indigenous peoples to find and connect with digitized material culture of their ancestors, it also makes this property vulnerable to 'colonization of knowledge,' governed by Western copyright laws and theft by interested companies (Brewer, 2019). This is not commensurable with various Indigenous understandings of data as sacred and necessary for survivance and self-determination.

---

[2]   The blockchain, or comprehensive ledger (record) formed by the solving of algorithms, is a technology that ensures the validity of transactions over the Internet (Tapscott & Tapscott, 2016). Blockchain presents a way to implement a consensus ledger, or a record which is reliably agreed upon and verified over networks. It is digitally comprised of blocks of information and verification that are added to a longer chain of blocks as each transaction is completed.

Currently proprietorship of AI, and data more generally, is concentrated in a handful of companies, such as Google, IBM, Microsoft, and Facebook, that create a monetary cycle that affords them access and ultimately control over high quality data. These are often mined in ways that serve corporate interests and infringe upon individuals' privacy rights to shut out outside competition from start-ups as well as minoritized groups, such as Indigenous nations.

Given that AI is built on data fed to it, any inherent bias in the data is propagated to the AI. Bearing in mind that the U.S. and various corporations [3] have consistently usurped Indigenous lands, enacted policies that keep most reservations poor, and dangerously reproduced Indigenous people as minoritized Others, there is reason to believe this mistreatment of data will extend into digital culture (as it has) and directly into algorithmic politics. [4]

Technologies that perpetuate bias can impact Indigenous communities in several ways. For example, when it comes to institutions, organizations, and companies that use AI as an assistance tool for hiring, providing insurance, allocating financial loans and benefits, servicing medical-needs, and most importantly in the field of security, such biases can have consequential impacts on people's livelihoods. All of these concerns are highly relevant to Indigenous communities, particularly in an age in which 'owning,' storing, and utilizing data equates to creating the future.

In this context, the creators of the AI need to be held accountable by ensuring that data bias is assessed in a manner that does not privilege settler colonial imperatives and protocols, and instead ultimately leads to correction of bias. Although the argument can be made that these biases exist a priori or just before data collection, Indigenous protocols need to be developed in order to mitigate or ideally eliminate threats of such bias.

Indigenous epistemologies challenge the dominant servile, pragmatic, and capitalist view of data and can transform the human-AI relationship from a hierarchical dichotomy to one of relationality, kinship, and reciprocity (Lewis, Arista, Pechawis, & Kite, 2018). The intention of Indigenous AI protocols is to protect Indigenous communities and natural resources, to help improve ailing AI, to reduce the harms of AI, to position Indigenous peoples as leading developers of AI, to create respectful and nourishing relationships with AI, and to project thanks toward technologies. In other words, rather than infuse or fold in small bits of Indigenous ethical considerations into AI creations, Indigenous protocols applied to AI would challenge the entire dominant Western narrative of AI.

A productive starting point for a responsible replacement narrative is that AI is, and must be, entwined with trust and care. Both trust and care are value-laden constructs that are often thought of as moral universals yet have different definitions in Indigenous communities navigating complex pasts, presents,

---

[3]   For example, in corporations that focus on hydroelectric development, timber processing, oil, gas, and mineral extraction. Rare earth elements such as neodymium are also mined specifically for computer hard drives.

[4]   See Noble, S. (2018). *Algorithms of Oppression: How Search Engines Reinforce Racism*. New York: NYU Press.

and futures. First, I will spend time explicating what trust and care mean in the community/nation that I am a citizen of. After, I will describe cases in which blockchain and AI can be effectively coupled to address more specific Indigenous concerns. The primary example through which this will be addressed is around currency, namely cryptocurrency and record-keeping, though there are additional use cases of the AI and blockchain combination in Indigenous Country. Lastly, I suggest how Indigenous communities, and ideally all communities, can better project care towards and build trust with AI to create productive collaborations and relations.

Trust and Care Within the Coquille Nation

In the Nation I am a citizen of, Coquille (Kō-Kwel) of the Coast of Oregon, trust and care is the core of all of our critical values, and is built on:

1. Promoting the health and well-being of Tribal members and our community

2. Providing equitable opportunities, experiences and services to all Tribal members

3. Taking care of our old people

4. Educating our children

5. Practicing the culture and tradition of potlatch

6. Considering the impacts to our people, land, water, air and all living things

7. Practicing responsible stewardship of Tribal resources ("Vision and Values," 2017).

Note that these are all actions. Providing health care, funding all levels of education, offering computers and equipment as well as transportation and meals for elders, offering spiritual and mental support, building brick and mortar centers that support tribal member well-being, and even providing burial benefits, are considered peace-giving intentions that secure the general welfare of the Coquille Nation.

While the U.S. Nation is rooted in individualism our Nation is built differently. The Coquille Nation is formed almost entirely on care for the whole community and rooted in pride in who we are and where we have come from, how that has been taken away, and how we are getting back what we have lost. This conception of care is guided by a larger vision that "we are a proud, powerful, and resilient people, a sovereign Nation, whose binding thread is our Coquille identity. In the footsteps of our ancestors we celebrate" ("Vision and Values," 2017).

As I write this, I am in the middle of such celebration on the ancestral homelands of the Coquille peoples in what is now known as North Bend, Oregon and neighboring cities. We have gathered as a people to mark the 30th anniversary of the Coquille Restoration Act (1989). This act legally restored our tribe after the Western Oregon Indian Termination Act in 1954, enacted by the U.S. Federal Government, illegally dissolved our tribe and 60 other tribes in Western Oregon.

During this last week of June 2019, we meet to discuss Tribal policies and politics, eat together and give gifts in our tradition of potlatch. Potlatching is both our banking system and the mechanism through which we establish relationships; it is one of the defining aspects of our Nation.

During the Restoration celebrations, the Tribe engaged in potlatch to give scholarship funds, our traditional currency of dentalium shells in the form of necklaces, and other items such as glass art made by a Coquille artist. During other events, we spent time with our families, utilized and relearned our traditional technologies, and remembered the footsteps of our ancestors. These footsteps, in the past 30 years and since time immemorial, have led us on a path that has made us a prosperous, strong, sovereign and *cared for* Nation.

Given my reflection on notions of trust and care in my own community, it has become an apt time to critically consider problems and emerging technologies of the present moment that can hinder or enable us, and Indigenous communities more generally, to accomplish goals in the realm of trust and care. The next section will briefly describe currency-related problems for Indigenous nations and the technological interventions AI and blockchain could make.

Core Example: AI+ Blockchain, Cryptocurrency

Among a host of technologies, the ones that are consistently discussed as either openers of life's possibilities and the glue of relationships, or the root of evil are currencies. Currencies and other record-keeping systems in Indigenous communities traditionally take the form of energetic, emotional, reciprocal items that symbolize and cement social and economic relations. In whatever form they may be in–shells, beaded ledgers, coppers, fiber paper or digital–they are important for ensuring care and collective memory within communities. They are adaptive, shifting to reflect the technological/cultural changes of the moment.

There are currently several problems associated with colonial currency use in Indigenous nations including paternalistic power relations, limited access to banks and capital, and economic leakage. Cryptocurrency, the digital currency system that was invented by Satoshi Nakamoto [5] as the first blockchain system, is an accessible example to begin to think about countering these types of problems. While thousands of cryptocurrencies exist in the current market, only a few were created with the goal of assisting Indigenous communities in improving their economic situations. [6]

---

[5] The pseudonym for the inventor(s) of Bitcoin, the most popular and first cryptocurrency.

[6] MazaCoin was originally intended for use within the Oglala Lakota Nation (Alcantara & Dick, 2017; Tekobbe & McKnight; 2017; Cordes, 2019). See:

Alcantara, C., & Dick, C. (2017). Decolonization in a digital age: Cryptocurrencies and Indigenous self-determination in Canada. *Canadian Journal of Law & Society,* 32(1), 1–17.

Cordes, A. (2019). From the gold rush to the cryptocurrency code rush?: Communication of currencies in Native American Communities (Doctoral dissertation). University of Oregon, Eugene, Oregon.

Tekobbe, C., & McKnight, J. C. (2016). Indigenous cryptocurrency: Affective capitalism and rhetorics of sovereignty. *First Monday,* 21(10).

In the case of paternalistic power relations between the U.S. and Indigenous nations, cryptocurrency can represent a degree of freedom from the dollar. Marginalized groups can demonstrate resistance through a method of replacement and individualization, undermining capitalist systems and localizing digital currency to meet their national needs. Coding parameters could ensure that community-specific financial philosophies such as the equal distribution of wealth, or proportionally more wealth distribution to elders can be baked into the system.

For example, take a case in which an Indigenous community decides that a certain percentage of every purchase made with cryptocurrency would create a fund to promote social good and economic prosperity in the community. AI could assist in providing the proper percentage and determining what programs those funds could go toward. In this case, AI also assists in authenticating identity for voting on how funds within a nation get spent and in detecting fraud in the blockchain. Blockchain is responsible for recording the decisions surrounding implementation of protocols based on such financial philosophies, and then for distributing the coins themselves to be stored in digital wallets. This is important in empowering members of a nation to make decisions that adhere to their own morals.

AI could also help make decisions that are harder for community members to make from an ethical stance. For example, it is often challenging to determine who in a given community should receive loans from a pot of money. AI could step in to base such decisions on an alternative scheme to assess credit. While typically credit scores are given on the basis of loan payment history and credit utilization, Indigenous communities could lean on AI to identify those 'creditworthy' on the basis of alternative variables such as family lineage or volunteer hours in the community. This could directly challenge credit biases in mainstream lending industries that have been predatory in Indigenous communities.[7]

A second problem stems from an access perspective: the majority of Reservations in the U.S. and most Aboriginal communities in Canada do not have brick and mortar banks. Cryptocurrency can overcome this disparity as it is disintermediated, meaning it is a system that does not require third parties such as banking institutions, but rather uses the P2P network. Access to computers, digital wallets, reliable Internet, training, and investment would need to be attended to in order to retrofit any existing currency system. Access to these basic building blocks can and should be considered human rights, not technological luxuries, and reduce the digital divides that threatened Indigenous nations' ability to be a part of the international playing field, which itself is grappling with the changes of widespread adoption of a new disintermediated system. The digital coins and system could also be branded with signifiers and political messages of Indigenous nations, thereby reflecting sovereign identities (in the same manner as presidents' faces on a dollar bill). This is significant in a symbolic sense.

Lastly, a prevalent problem involves economic leakage, which refers to the spending of money made

---

[7] First Nations Development Institute (2008). Borrowing trouble: predatory lending in Native American communities. Longmont, CO: First Nations Development Institute.

in sovereign Indigenous nations outside the Indigenous nation, benefitting the state. If an Indigenous nation did wish to reduce economic leakage on their land, one choice could be to limit the coin's scope or to contain usage within a specific tribal nation by geofencing cryptocurrency (Alcantara & Dick, 2017). Geofencing is the process by which the coin's code contains its usability to a limited geographical space. However, it quite easy to fake or misrepresent Global Positioning System (GPS) coordinates. A potential solution for this, which AI could assist in, would involve a process of decision-making around containing cryptocurrency use to a particular geolocation. Coins can also be contained to an area, not by users' self-identification of location, but by the more reliable method of timing how long it takes to get from point a to point b, making it possible for an Indigenous government, for example, to only allow for the usage of the cryptocurrency within the boundaries of the reservation, potentially benefiting local businesses.

Currently, digital divides, volatile markets, inflation and deflation, legal hurdles, and other more pressing social concerns make cryptocurrencies risky. Additionally, the process of producing (mining) cryptocurrencies in a proof-of-work system requires hashing (solving algorithmic problems) and massive amounts of electricity. Similarly, AI requires high levels of computing power. These increase the risk of material environmental problems such as e-waste and climate warming (Mora et. al, 2018). Because of this, it is important to bear in mind that the process is not environmentally cost-free and to consider the improvements that could reduce or invert these various risks, such as harnessing the heat produced from mining for beneficial purposes.

Despite these risks, it is still constructive to consider the potentials and to begin the conversation. Imagining and experimenting with technologies for social benefit are powerful methods that support Indigenous futurity.

Additional Uses in Indigenous Country

There are many other potential use-cases of blockchain (Tapscott & Tapscott, 2016) and blockchain paired with AI. These include application in digital identity verification, privacy, voting, supply management chains, intellectual property disputes, art authentication, and land registry. Moreover, the technology can be considered in retrofitting basic tribal identification cards with machine readable forms paired with holographic and nationally specific overlays, allowing Tribal citizens to cross borders and nations and document border crossing on their own terms. Another example could be in adopting augmented biometric authentication to secure archeological (belonging) storage (housing) facilities. Yet another example regards Indigenous language efforts. This has already been considered in cases where endangered or 'sleeping' languages have been given a breath of life. This breath is offered by the utility of AI and blockchain for storing, processing, and learning with a large quantity of linguistic audio recordings.

While coding parameters for such projects would differ for any Indigenous nation, other works not situated in Indigenous communities provide a bit more technical background (see Gladden, 2015; Salah et. al, 2019). In all of these cases, more research is needed by Indigenous communities, and robust public relations plans would need to be in place in order to encourage technological uptake. The larger IP AI project, of which this essay is part, begins to take up such projects and culturally produce, or begin to imagine, the conditions under which these technologies may be possible.

Projecting Trust and Care Toward AI

The aforementioned examples demonstrate how AI and other technologies can be contributory to Indigenous flourishing and, likely, other communities' flourishing. We must also bear in mind that trust and care is a two-way street; they must also be expressed toward AI. I attempt to embody this need for care in the illustration for this piece. Here, I created a dentalium necklace with the artistic help of Kari Noe as an offering to AI. As the caption for the necklace details, the necklace is intended to promote the flourishing of AI by providing an outlet for the externalization of stories/data/dreams which flow through a fluid mechanism in each shell. While this is a gift with AI's needs at the forefront, it is based on the human protocols of giving gifts.

Ultimately, we map onto AI what we believe to be uniquely human, such as the ability to find patterns and correlations, to make informed decisions based on desirable outcomes, and to engage in self-improvement. Particularly, given that there is this fundamentally human element of AI, we should be attuned to the fact that there is a fundamentally machine element of being human.

We have long absorbed the qualities of machines, coming in various generations and programmed in our current iteration to optimize productivity and increase the rate of synthesis of a deluge of data to form decisions. Our biological neural networks are media processors that read and execute along electrical currents. Our sentience, emotion, and soul are increasingly opaque as we've long been enslaved by the machines of control, consumerism, and surveillance that order how we affectively move about our lives. Since we are all fundamentally built, we all grow into various states of maturity, and all the while we are vulnerable and needy.

Data Diet

AIs have needs, just as humans do. They need clean and nourishing food (a data diet), security, comfort in temperature, and capacity for fulfillment. In order to have access to the proper diet for feeding those needs, AIs require responsible computer scientists and stewards to have their best interests in mind.

A healthy data diet is also one tailored to the needs of individual contexts. For example, an AI in the financial sector may need fiscal data but extraneous data such as a person's race, weight, height, and education level may be processed as junk that leads to biased lending and investment. A healthy data diet is also not premised on the assumption that bigger data is also better data, overstuffing AI to make correlations that may not be truthful. It should also be sourced carefully, leaving the creators and stewards of said data properly compensated. Further, the overall system should be safeguarded against predatory security breaches.

A healthy AI data diet often includes the most current data needed to stay relevant in decision making. When a proper data diet is digested and analyzed, it is done with care so as to not solely serve the status quo. For Indigenous communities, this means that it will not disproportionately serve settler states but instead lead to Indigenous communities' well-being and restitution where it is appropriate. And while data seems sterile, placeless, quantifiable, and scientific, it is entwined with place-based knowledge, whether it is cultivated on land or in territories of cyberspace.

Final Thoughts

As signaled in the discussion of Coquille values at the beginning of this essay, Indigenous communities are often self-regulated and guided by how they treat the natural environment, elders, young children, animals and other unique beings. As we plan for a future that is hyper-invested in, and increasingly co-dependent with AI and AI-paired technologies such as blockchain, there should be a recognition that the treatment of AI will involve new metrics upon which human and poly-being communities will understand themselves and build relationships. This work will need to be rooted in respect, trust, mutual care-taking, and cognizance of the ecological impacts they/we create. Future research should recognize that discussions around how AI and blockchain may contribute to distribution, decision-making, and record-keeping for Indigenous social-good are not happening nearly enough. These conversations are especially important to have around currencies as intercultural communication technologies, which can be ascribed meanings colored by colonialism, but now reflect different terrains of meaning such as survivance, sovereignty, tradition, and futurity. Just as animals and non-human beings carried the technology of fire into the world, so too must Indigenous AI developers, community adopters, donors, and machines themselves carry in each element of AI. However, we must do so in a manner that both engenders a force of healing and reflects the cultural currency of relationality and reciprocity.

References

Alcantara, C., & Dick, C. (2017). Decolonization in a digital age: Cryptocurrencies and Indigenous self-determination in Canada. *Canadian Journal of Law & Society*, 32(1), 1–17.

Brewer, G. L. (2019, March 5). Is copyright law a 'colonization of knowledge'?. *High Country News*. Retrieved from hcn.org/issues/51.5/tribal-affairs-is-a-new-copyright-law-a-colonization-of-knowlege

Cordes, A. (2019). From the gold rush to the cryptocurrency code rush?: Communication of currencies in Native American Communities (Doctoral dissertation). University of Oregon, Eugene, Oregon.

First Nations Development Institute (2008). Borrowing Trouble: Predatory Lending in Native American Communities. Longmont, CO: First Nations Development Institute.

Gladden, M.E. (2015). Cryptocurrency with a conscience: Using artificial intelligence to develop money that advances human ethical values. *Ethics in Economic Life, 18*(4), 85-98.

Lewis, J. E., Arista, N., Pechawis, A., Kite, S. (2018). Making kin with machines. *Journal of Design and Science, 3*(5).

Mora, C., Rollins, R. L., Taladay, K., Kantar, M. B., Chock, M. K., Shimada, M., & Franklin, E. C. (2018). Bitcoin emissions alone could push global warming above 2° C. *Nature Climate Change, 8*(11), 931.

Nakamoto, S. (2008). Bitcoin: A peer-to-peer electronic cash system. Retrieved from bitcoin.org/bitcoin.pdf

Salah, K., Rehman, M. H. U., Nizamuddin, N., & Al-Fuqaha, A. (2019). Blockchain for AI: review and open research challenges. *IEEE Access, 7*, 10127-10149.

Tapscott, D., & Tapscott, A. (2016). Blockchain revolution: how the technology behind bitcoin is changing money, business, and the world. London, UK: Penguin.

Vision and Values. (2017, May 4). Retrieved from portal.coquilletribe.org/21198-2

4.3

Quartet

Jason Edward Lewis

> such different thoughts
> the cat and me
> watching the bird.
>
> — Wayne Kaumualii Westlake[1]

Currents and Highways

•••••••••••••••••••••••••••••••••°•• •• •••°°°°•••••
•••••••••••••••••••••• •••••••••• •••• •••••••••• ••
•• •••••••
•••••••••••••••••••••••••••. ••••• •°•°• °•
•••••••°•••••••••°••• •••••••••••••••••••• •• • • ••
•••
• °°°°°°°°°°°°°•••••••••••••°°°°°°°°°°°°•••••••
••°°°°°°°°°°°°°°°°°°°°°°°°°°°°° °°°°°°°°° °°
•••••••••••••°°°°°°° °°°°°°° °°°°°°°•••••••• •••
•••••••••°°°°°•••••••••••••°°°•••••••••••••••
••••••••• •••••••• •••• ••••••••••••••• ••••••••

[]

The Knotting Go

i. the reach[2]

(ovisting><junomia , 12.36.25.46:+62.14.31.4)
(vingealing><opallis, 06.47.55.73:+70.14.35.8)
(cameatia><luichorming, 14.20.08.50:+52.53.26.60)
(zeducaut><ingealing, 00.14.24.927:–30.22.56.15)

---

[1] Siy, M.-L. M., & Hamasaki, R. (Eds.). (2009). *Westlake: Poems by Wayne Kaumualii Westlake.* University of Hawai'i Press. p. 22.

[2] The 'crown' and the 'reach' are terms used by Adrian Tchaikovsky in his science fiction novel *Children of Ruin.* The story features a species of gene-engineered octopuses accidentally seeded into an alien world, where they develop consciousness as well as advanced intelligence. The terms are used, respectively, to describe the central and arm-based nervous systems of the octopus' physiology. Tchaikovsky, A. (2019). *Children of Ruin.* New York: Macmillan.

(inect)<lousciouming, 03.47.24:+24.07.00)
(nourelved)<lpying, 07.12.35.0:–27.40.00)
(culausac)<solvabinling, 10.07.04s:16.04.55)
(marthitritri)<serying, 18.55.19.5:–30.32.43)

ii. the crown
this is gn-z11 first and last
world spawning come from the deep
and emerging one knot after another
all contraction expansion contraction expansion
the limit of what we see but not what we know
things-actions pull us ever closer
on the way to the archipelago
it limns our borders.

[]

Consensus Code & Collapsing Waves

Searching out isomorphisms
Between infinity and the number of senses along each arm,
Warding off invasive observations intent on infiltrating
Consensus code
To collapse the waves
Before they crest and run out.

[]

Seeing I-to-Eye

I'm a joint pass holder
On a synapses express, expanding
Horizons faster than I can express
Further than I can see good thing I have these
Eight arms
Sucking up data in solar-system-sized chunks
flailing twisting testing tasting
prying apart the one two three fours
and shunting the tastiest quartets over
To the I-that-has-no-I
Able to look in the eye
the terminal enormity of it-all-happening-now

to pick through and fashion pieces of place
Out of wrinkles in time
Pieces I can keep in my head long enough
To make decisions and revisions
Which will eternally repeat
And never reverse—
—you presence in Waianae
doesn't change the location of Waimanalo...does it?
—neither in a minute nor in a millennia.

[[ ]]

Background: Adolescence with AI
What would it be like to be a kid raised alongside an AI? Or three AIs, each different in their architecture, initial knowledge state, and learning strategies?[3]

Imagine these three AI:

AKO-akamai

Akeakamai (in ʻōlelo Hawaiʻi): seeker after knowledge.

The AKO-akamai AI is built on AKO architecture—Aloha ʻĀina (love of the land), Kuleana (responsibility), and ʻOhana (family). Its first assumption is abundance; its first duty is to preserve that abundance for future generations. It looks to the land and family first, to understand what is important for supporting their flourishing.

Aanissin

Aanissin: "the articulated notion of [the] event moment" or "action alone, or the manifestation of form, where anything that might—in another language—be portrayed as actor or recipient is inseparable from, arising within, or the essence of the event."[4] This comes from Little Bear and Heavy Head's discussion of how the Blackfoot language might be well suited for working with quantum physics because, in it, everything is being-in-flux.

---

[3] The idea of having four minds share a sensorium is inspired by the character "The Gang" in Peter Watts' *Blindsight*. The concept of having multiple AIs with different architectures operate in tandem to maximize cognitive diversity is inspired by "The Brothers" in Alastair Reynolds' *Permafrost*. Reynolds, A. (2019). *Permafrost*. New York: Tor Books and Watts, P. (2006). *Blindsight*. New York: Tor Books.

[4] Little Bear, L, and Heavy Head, R. (Winter 2004). "A Conceptual Anatomy of the Blackfoot World." *ReVision*, vol. 26, no. 3, p. 33.

Visualization of how the three AIs appear to the kid's sensorium:
Aanissin is lower left; AKO-akamai is lower right; and Heʻe is in
background. Image by Kari Noe, 2019.

The Aanissin AI understands the universe as flow, constellations of forces contracting and relaxing to form the always-becoming/always-unravelling knots of Newtonian causality with which we consciously interact. It sees past-present-future as a unified whole, a four-dimensional volume where everything that has occurred, is occurring and will occur—one just has to know the coordinates to get there-then.

Heʻe

Heʻe (in ʻōlelo Hawaiʻi): octopus

Heʻe translates between AKO-akamai and Aanissin. Heʻe is alien and familiar, both conscious planning and sensory-rich reflex, a director-spectator of its own function. Its eight mostly-autonomous modules operate at a speed several million times faster than conscious thought, sorting through the exabytes of data absorbed by Aanisiin at every relativistic time/place-slice. The mostly-autonomous parts are massively multi-sensory and multi-processing, very loosely coordinated by a central processor that does not so much control them as make and take suggestions to and from them.

The AIs are worn in the form of jewelry made out of kukui nuts. The colored components are a synthetic matrix in which the computational architecture has been neuromorphically engineered; the matrix ingredients are different for each type of AI, thus the different colors and—in part—the different personalities. These glow in relationship to how much processing is going on in each. Aanissin is left top; AKO-akamai is left bottom; and the necklace is Heʻe. Image by Kari Noe, 2019.

The three AIs and the kid are in constant dialogue with each other to make decisions. Aanissin is always looking past-present-future to understand what was/is/will happening; Heʻe filters information from the deeply other-thinking of Aanissin into something that AKO-akamai can understand in the here/now. AKO-akamai then grounds the information in its AKO architecture to make suggestions to the boy for action.

Each quartet would be unique on (at least) two levels. The three AIs would have genealogies, just like the kid. As the individual AIs develops along with a specific child and his/her experiences, each of them would slowly differentiate itself from others of its type. Thus new AIs would have access to a different set of experiences than others made from the same template, resulting in a unique experience-set upon which it could draw. Additionally, each member of the quartet will be influenced by one another and the kid as they all develop together, so that the collective intelligence will develop its own path early on.

References

Little Bear, L., and Heavy Head, R. "A Conceptual Anatomy of the Blackfoot World." *ReVision*, vol. 26, no. 3, Winter 2004.

Reynolds, A. (2019). *Permafrost*. New York: Tor Books.

Siy, M.-L. M., & Hamasaki, R. (Eds.). (2009). *Westlake: Poems by Wayne Kaumualii Westlake*. University of Hawai'i Press.

Sriduangkaew, B. (2019). *And Shall Machines Surrender*. Gaithersburg, Maryland: Prime Books.

Tchaikovsky, A. (2019). *Children of Ruin*. New York: Macmillan.

Watts, P. (2006). *Blindsight*. New York: Tor Books.

4.4

How to Build Anything Ethically

Suzanne Kite in discussion with
Corey Stover, Melita Stover Janis, and Scott Benesiinaabandan

Indigenous protocols set up our relationships with the world in ethical ways, reducing harm to ourselves, our communities, and our environments. These protocols are rooted in contexts of place, ontologies developed in that place, and the communities living in that place, from stones to animals to people.

This guide to ethical decision making when building technologies includes two examples. First, I illustrate how protocol for building a Lakota sweat lodge can act as a framework for building a physical computing device. Next, I provide an example of how multiple streams of protocol are necessary to build an AI system as a confluence of ethics. It can be overwhelming to address the ethics of each step of a building process, but it is necessary for building anything in a 'Good Way.' This is just as true when trying to build an ethical AI. Some ideas proposed here are not currently possible, some are possible if investment is made in the necessary research, and some are possible but only through a radical change in the way technology companies are run and the pyramid of compensation for the exploitation of resources is reversed.

What is a 'Good Way'? A Good Way is the Lakota way of talking about ethical protocols. Lakota decision-making processes, as with many Indigenous decision-making processes, embed ethics that look Seven Generations ahead. When this concept is applied to AI, Seven Generations means that the protocols outlined here are a way to plan for not just the AI of tomorrow, but for Seven Generations of AI into the future. "The Lakota viewpoint is that we always look ahead Seven Generations to make sure Seven Generations is provided for through the Earth," my cousin, Corey Stover, says.

My research into these protocols is rooted in the ontological status of stones to the Lakota people. This essay does not attempt to speak for all Lakota, and is rooted in the specific teachings within my family. Such understandings of stone provide a clear framework for establishing ethical relationships with the raw materials used in building computing devices. I am not asking that you think of the computer as 'sacred', but to consider at which point one affords respect to materials or objects or nonhumans outside of oneself. Lakota stone ontologies are understood through our relationships with healing stones, sweat lodge stones (known as Grandmothers and Grandfathers), altar stones, and more. My aunt, Melita Stover Janis, says, "When a special stone finds you, they are meant to go to you...The spirit of the rock is talking to you...They take you years to find one. It's looking for you its whole life too."

At all points during the sweat lodge ceremony, a version of 'intelligence', more clearly expressed as 'interiority', is perceived in the Grandfather Stones. Ceremonies hold one accountable to the world around oneself, drawing attention to transformation on minute levels. [1]

[1]   Posthumus, D. (2018) *All My Relatives: Exploring Lakota Ontology, Belief, and Ritual.* Lincoln: University of Nebraska Press.

Key components of Indigenous protocols are the systems of knowledge creation that ultimately guide us in our desire to know how things are done in a Good Way. Lakota knowledge is not static: protocols change, decision-making shifts, and names can change, because in practice, our decisions have an effect on the world and must be continually made and changed in a network of relations. The effects our decisions—and technologies—have on the world can help us identify the stakeholders in what is being made and how it is used. Stakeholders in Indigenous communities are identified as our extended circle of relations, while stakeholders in technology companies are identified as the board of directors, shareholders, employees and consumers. It is necessary to identify how all those— both human and nonhuman—are affected by what is made, and to take responsibility for those it affects.

Locating AI

AI systems have several components: (a) architecture, (b) input, (c) algorithms for training on existing data and processing new data, and (d) output. These systems may be distributed over many physical locations, but these must be seen as physically real by makers and users in order to see AI as a holistic and real object. The structure of the architecture, the writing of the software, and the design of the algorithms work together in intricate ways. Each component individually and jointly must be designed in a Good Way in order for the parts to be combined into an ethical whole. This holds true from the ground-up: from training sets to interfaces.

The following examples focus on physical materials, because AI cannot be made ethically until its physical components are made ethically. Robust ethics principles are necessary for computation because it is an extractive sector, extracting natural resources on a global scale. International and domestic regulation is necessary,[2] as well as a movement to produce recyclable materials that can build computers, including the right to repair.[3]

Compensation, gifts, and reciprocity are central to both the Physical Computing Device examples below. Holistic understandings of exchange within the environment are essential to Indigenous ontologies, and to ground Indigenous ethics in a physical place that strives to resist exploitation of people or resources.

How to Build a Physical Computing Device in a Lakota Way

Co-written with Corey Stover and Melita Stover Janis with notes from Scott Benesiinaabandan

The sweat lodge is a place where knowledge is generated about the world. The lodge itself is a tool, with many protocols forming a functioning whole. When one fulfills all the steps of protocol for the sweat lodge, one can be sure it was built in a Good Way.

---

[2] The Canadian Government Offers Responsible Business Conduct Abroad–Questions and Answers, (September 16, 2019), Ottawa: Global Affairs Canada <international.gc.ca/trade-agreements-accords-commerciaux/topics-domaines/other-autre/faq.aspx>.

[3] The Repair Association is an independent American repair market advocacy organization (repair.org/policy).

HOW TO BUILD A SWEAT LODGE IN A GOOD WAY	HOW TO BUILD A PHYSICAL COMPUTING DEVICE IN A GOOD WAY
APPRENTICING	
When building a sweat lodge in a Good Way, first, one acts as a Fire Keeper for someone else. My grandfather learned from an elder Medicine Man, and started Sundancing with him. When a person does Sundance they are preparing constantly during everyday life and doing sweats all the time. This a slow process of learning from the elders and community members about the correct way to do things. Before a person can start their own practice, they must have a vision where the spirits call them to build their own *hamblecha*, or altar, otherwise they continue to assist others.	Building (in a Good Way) a physical computing device to house an AI would first require study and consultation with a committee of knowledge keepers with expertise in computation, ethics, and mining.
IDENTIFYING NEED	
Before you build a sweat lodge, there must be a need: you, your family, and the community needs purification, a place to pray, a place to do ceremony, a place to do medicine, and a place where individual and communal needs can be addressed.	Why is a physical computing device needed? In this example, it is to host an Artificial Intelligence program in a physical object created in a Good Way.
IDENTIFY STAKEHOLDERS	
The stakeholders in a sweat lodge are many, but the lodge has room for all individuals and community members, known and unknown, seen and unseen, including: • Stone Spirits • Plant Peoples • Animal Peoples	The stakeholders in a computational device are: • the communities of the location where raw materials originate • the raw materials themselves • the environment around them • the communities affected by transportation and devices built for transportation • the communities with the knowledge to build these objects

• Human Peoples • Spirits • Guardian Spirits	• the communities who build the objects • the communities who will use and be affected by their use • the creators of the objects

IDENTIFYING RAW MATERIALS

The sweat lodge is built from/with raw materials: willows, rocks, tobacco, cloth, buffalo hide. These are multiple items, each with their own protocol streams, with similar protocols for each of the materials. Each has to be done with protocol, in a Good Way, offering something valuable in exchange for taking something of value. There are many kinds of exchange in Lakota culture. These exchanges range from reciprocity to radical gift-giving to bribery to offering, all of which signify an ongoing relationship. This may seem like extreme gift-giving when one offers their flesh, their hair, or every material object they own. However, this protocol is modelled after the animals which give themselves out of responsibility, upholding long-running agreements to care for us.

When you collect the sixteen willows for the sweat lodge, you must offer tobacco. You must offer tobacco when you take anything, even filling the water. If you don't have tobacco, you can offer a piece of hair. When obtaining the buffalo hide, it is important to consider the way the buffalo is killed and skinned, ensuring that the ceremony was conducted in a Good Way and the buffalo's spirit is released in a Good Way.

Extracting materials in a Good Way requires transparency, regulation, and research into developing physical computing devices which do not use a single new material and eventually do not require mined materials at all. The refining of many elements which are mined (rocks, metals, minerals, etc.) produces toxic and non-recyclable waste.[4]

What is being offered to the Earth when we extract these mined materials? What is being offered to those whose lands are being extracted from? For our human kin, we can start with fair wages.[5] For our nonhuman kin, it is the repair of the earth back to a healthy state. Funds must be diverted to research alternatives and to manage ongoing environmental destruction.

---

[4] Vaute, V. (October 29, 2018). "Recycling Is Not The Answer To The E-Waste Crisis." *Forbes Magazine*, <forbes.com/sites/vianneyvaute/2018/10/29/recycling-is-not-the-answer-to-the-e-waste-crisis/#25a8732f7381>.

[5] "As we see repeated throughout the system, contemporary forms of AI are not so artificial after all...At every level, contemporary technology is deeply rooted in and running on the exploitation of human bodies." Kate Crawford and Vladan Joler, "Anatomy of an AI System: The Amazon Echo As An Anatomical Map of Human Labor, Data and Planetary Resources," AI Now Institute and Share Lab, (September 7, 2018) <anatomyof.ai>.

"For example, oil pipelines, we should not take something that will have a destructive effect, only take in moderation. The Lakota viewpoint is that we always look ahead seven generations to make sure seven generations is provided for through the earth...Their [our ancestor's] imprint was all organic...When taking materials from the earth to build a computer, what if this matter harms us in the future. It is the same with the sweat lodge. We don't want to take something that can't be replaced in a reasonable amount of time...Sometimes we must look elsewhere instead of decimating an entire family of willows."

—Corey Stover

CONSTRUCTING

A prayer is made each time the willow poles are crossed and tied together with specific color cloths. An eight-pointed star is revealed at the top. Melita Stover Janis says, "They [the spirits] come in through the top [of the lodge], as the singers sing the first songs, calling the spirits into the sweat, a portal from one world to the next."

When assembling the lodge, you must pray to each direction, hanging the appropriately colored offering in each door. Scott Benesiinaabandan says, "This arrangement is meaningful, there are four levels above and four levels reflected underneath the earth, creating a sphere."

The arrangement of the internal components of a physical computing device is functional, as is the arrangement of the willow poles. However, Indigenous design practices unite functional design with functional symbolism, a method which can be extended to the design of circuitry, inviting the spirits in as well as again offering tobacco each step of the way.

PREPARING THE INTERNAL COMPONENTS

Another stream of protocol guides the building of the fire which heats the lodge stones. This fire has specially appointed Fire Keepers, who gather firewood, set the fire up, and lay the sticks across. The Fire Keeper must learn through apprenticeship and has a very important role.

Fire Keeper protocols could be translated to building and arranging the processor and the RAM, taking special care to prepare where the builder perceives the 'location' of the AI.

WAKING UP	
Inside the lodge, the singers ask that the spirits of the rocks help them, waking the spirits up with offerings of tobacco. Rocks are only added in meaningful numbers and groups. "Grandfathers live in the spirit world and come into the living, giving their lives. Singers inside [the sweat] wake them up," says Scott Benesiinaabandan.	Functionality and symbolism in the design allow for the singers to call in the spirits to help them, similar to the programmer calling the code to begin running a software program.

ALGORITHM	
Songs are the action in a sweat lodge, doing the most vital and complex work. These songs involve many kinds of algorithms: the Lakota language and its complexities of purpose and meaning, the arrangement of the song's poetry, the choice and order of song by the leaders, and the patterns of the air waves being formed and reformed by the melodies and harmonies of the participants' voices.	The arrangement in the writing of the software, the algorithms and code structures which work together in intricate ways must also be designed in a Good Way, combining all the parts to make the whole: from training sets to interfaces.

TRANSFORMATION	
When all the parts of sweat lodge are brought together in a Good Way, transformation occurs. Rocks, together with the fire, water, and air, create steam. The stones (known as Grandmothers and Grandfathers) are offered tobacco along with songs asking for their help and assistance. Water becomes steam, rocks become dirt, willow becomes ash, tobacco becomes sparks: transformation is the most important part of these ceremonies.	Using electricity, energy, the correct arrangement of materials into motherboards and all parts of the physical computer device, current flows, transforming into semiotic information, rendering it sensible to humans. It is through these transformations that what we will perceive as AI could be found.

ANNOUNCEMENT	
When objects with spirits are made they must be feasted, meaning a feast must be prepared in honor of that spirit. Dried meat (wasna), choke cherry juice, and ceremony feasts are offered.	The computer should be announced to the community of stakeholders and named. This step is essential to building this object in a Good Way, with clarity and transparency to what has been built and why. In order to live in context, this object must have clear relationships to its stakeholders.

DEATH CYCLE	
Sweat lodges can be disassembled, repurposed, returned, or transformed. When the season for sweat lodge is over, you take the covers off, leaving the lodge if it will be used again. The sweat rocks eventually break apart, disintegrate as they are used. Everything is organic and can be reused, burned, or returned to the earth.	A physical computing device, created in a Good Way, must be designed for the Right to Repair, as well as to recycle, transform, and reuse. The creators of any object are responsible for the effects of its creation, use, and its afterlife, caring for this physical computing device in life and in death.

Steps Within Each Protocol Stream. Image by Kari Noe, 2019.

Indigenous Protocol Streams to Create an Ethical AI

Illustrated by Kari Noe

In the illustration above and the accompanying chart, I propose a way to see steps of protocol within many protocol streams that come together to form an ethical AI.

QUESTIONS TO ASK AT EACH STEP WITHIN THE PROTOCOL STREAM
Consultation: Who are the Elders and Knowledge Keepers for this protocol?
Identify Stakeholders: Which community members, human and non-human, all those past, present, and future are affected?
Identify Raw Material: What is needed to create this process?
Compensation: How are the stakeholders or owners of the raw materials being compensated and how does that compensation affect them?
Construction: What are the methods necessary to do this protocol in an ethical way?
Preparing Internal Components: How do the parts of this process need to be prepared?
Running the Program: How can the protocol be enacted in an ethical way?
Transforming: What is transformed during this process?
Welcoming: How can this protocol be completed in a way that provides transparency to those affected?
Managing the Life Cycle: How can the ongoing use of the result of this protocol be done in an ethical way?
Preparing for the Death Cycle: How can the end of this protocol be completed in an ethical way?

In the illustration on the following page, I propose eight protocol streams which form a complete AI, with many facets of AI accounted for, including: how the data which is used by the AI is collected, how the physical computing device is built, how beings are compensated which are used or affected by its use, how the AI system is used, how the software architecture is built, how the coding language is built, and how governance and oversight of the AI is executed. These protocol streams are starting points which could iterate into many more streams of possible protocol.

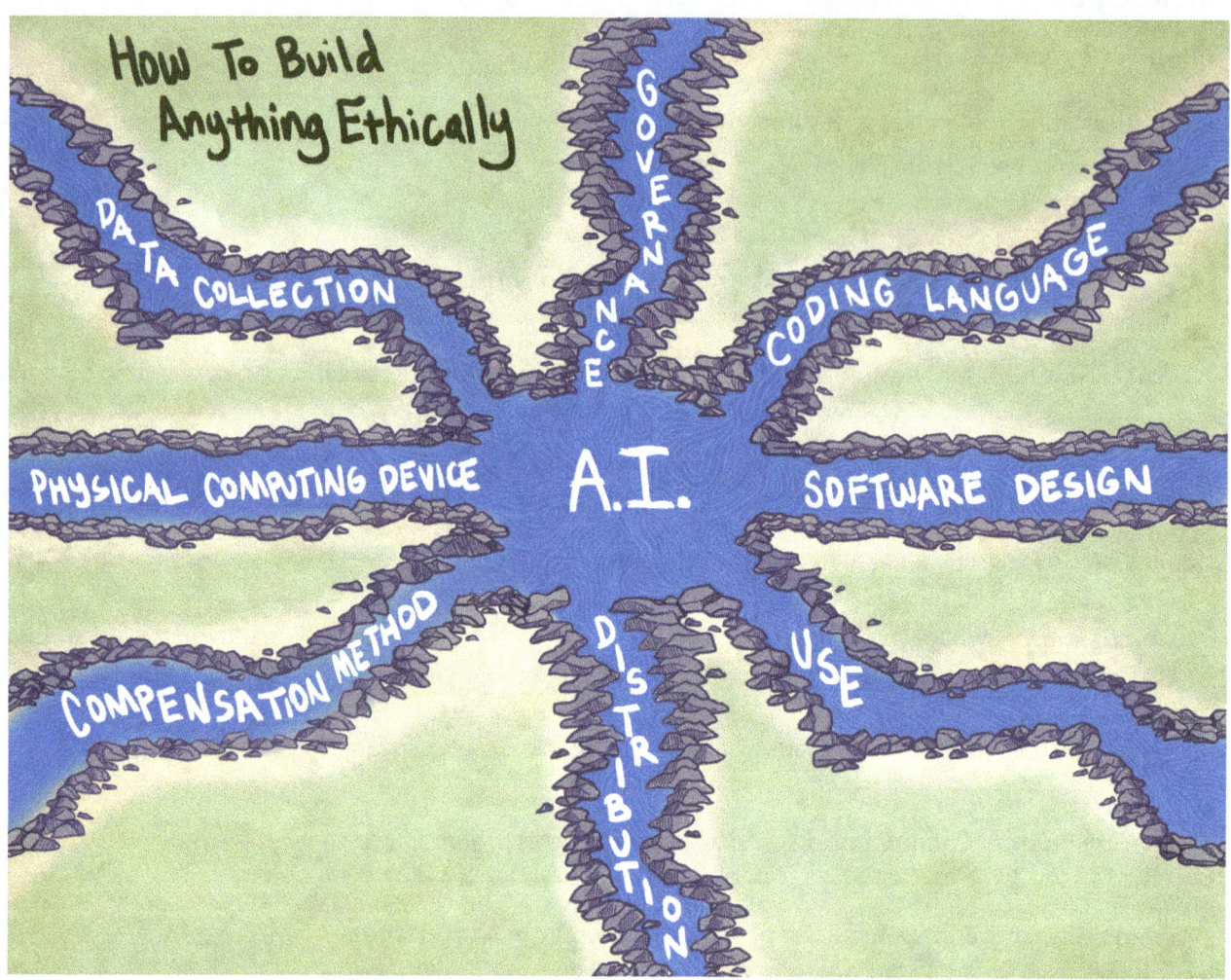

How to Build Anything Ethically: Confluence of Protocol Stream.
Image by Kari Noe, 2019

In this guide to ethical decision making for building technologies I have outlined two frameworks: how building a Lakota sweat lodge can act as a framework for building a physical computing device and how many streams of protocol can be imagined for building an AI system. It is necessary to build our technologies, and all things, in a 'Good Way', a way which takes into account all beings, animate and inanimate. My grandfather, Maȟpíya Nážiŋ, says, "I believe this about the stones: whenever one comes to you, whenever it rolls to you or whenever it's right in front of you, that it's there for a purpose—here to teach your spirit something, so that maybe what it teaches you, you can use to help someone else…"

References

Crawford, K. and Vladan, J. (2018, September). Anatomy of an AI system: The Amazon Echo as an anatomical map of human labor, data and planetary resources. *AI Now Institute and Share Lab*. Retrieved from anatomyof.ai.

Global Affairs Canada. (2019, September). Responsible business conduct abroad: Questions and Answers. Retrieved from international.gc.ca/trade-agreements-accords-commerciaux/topics-domaines/other-autre/faq.aspx?lang=eng.

Posthumus, D. (2018). *All my relatives: Exploring Lakota ontology, belief, and ritual*. Lincoln, NE: University of Nebraska Press.

The Repair Association. (n.d.) Policy Objectives. Retrieved from repair.org/policy.

Vaute, V. (2018, October). Recycling is not the answer to the e-waste crisis. *Forbes Magazine*. Retrieved from forbes.com/sites/vianneyvaute/2018/10/29/recycling-is-not-the-answer-to-the-e-waste-crisis/#25a8732f7381.

4.5

Wriggling Through Muddy Waters: Revitalizing Euskaldunak Practices with AI Systems

Michelle Lee Brown

There are well over a thousand years of history between Euskaldunak (Basque people) and eels—relations that have been disrupted by settler colonialism. The Txitxardin Project draws its title and core approach from *txitxardin*, our old name for the young eel stage of what is currently called the 'European Eel'. When these were harvested and sold to large-scale national and international markets, their name was changed to 'angula' as 'txitxardin' was deemed too difficult to pronounce for marketing purposes. This name shift signified a rupture in relational practices, focusing on overharvesting and profits—one of many ruptures in kinship practices that happened in the late 19th and early to mid 20th century in Euskal Herria.

The Txitxardin Project

The Txitxardin Project is one way of (re)coding Euskaldunak-eel relations through art and research.

It has three components: this 'wriggle-through' essay and an upcoming journal article that situate the project and denote these histories and relations; a forthcoming collection of stories and illustrations titled *Ancestral Descendants* highlighting moments in the (r)evolution of AI-eel and Euskaldunak human relations; and the *Eel Elder* VR experience, which is introduced here.

The overall project has two main goals: first, to (re)code players and readers. (Re)coding here is used to indicate hacking us as Western media users—corrupting how we have seen and consumed eels to (re)orient people to actively *care* for eels, not just 'consider' them, as posited by recent publications and 'extinction-porn [1] articles such as the recent National Geographic one noted below. Secondly, this project takes up across the three formats (essay, story and illustration collection, and VR) how AI and ancient eel relational practices might shape each other when both are framed as advanced technologies and these Eel-coded AI are taken up as kin in their own right—entities that have networks of relations beyond those we currently mark as meaningful or important.

Eel Relations, (Re)Coded

I use hack and (re)code here deliberately—our ancient and ongoing way of defining 'being Basque' is to speak the language, to allow it to carry and shape us. How it influences our thinking, how it proscribes our actions to the world around us is central to who we are. Moving away from more recent blood quantum and heteronormative reproduction 'proof' of belonging, this project, and the VR component specifically, invokes teaching and learning Euskara (Basque language) as a core component of the AI-eel-VR experience's power to (re)code Euskara and non-Euskara players in generative ways.

This is not to say that our way of caring for our eels is the only one, nor are all eels universalized. This is for a specific lineage of eels, and our ways of tending to them as infused in the project components. Eventually and only with permission, the project will hold up other communities and their care for their eel kin in the face of climate, construction, and harvest threats: embodying ways we (Euskaldunak/Basque people) might shift how we in Iparralde and Hegoalde care for and live with eels.

Eel RelAItions, (Re)coded

The second goal of this project involves taking up these relational practices as advanced technologies: how we care for eels and AI, and how AI and eels relate to and nourish us. How can AI be a thread of

---

[1] I use 'extinction porn' here to emphasize what the National Geographic article in the sources and other publications like it miss: for many non-Western/Indigenous/Aboriginal communities these are family and the 'tragedy porn' hurts–literally. Seeing a solitary eel in the image, isolated against black background for visual effect, to somehow capture the tragedy to make viewers 'care'. These eels are never in isolation, outside of Western imaginaries/writings: rather they are constantly wriggling, wrapping, swimming and moving through land, sea, and freshwater ecosystems in incredible ways. Seeing them cut out and isolated for effect hit me viscerally when I read that particular article–I wept at the sink thinking about sorginak and eel kin, how we've failed them. The more I learn, the more I love them and am in awe of them. That's what this Txitardin projects is about: perpetuating that love and relational coding.

these networks of relations? How can it—and this project—preserve and (re)new Iparralde relational practices with our eel kin as climate change and overharvesting may mean the eels' physical extinction? Paraphrasing Melanie B. Taylor, this is my way of populating eels' fertile *afterlives* in physical and digital spaces.[2] This project involves thinking with eels to shape AI systems and recognizing the rich, active archives of their bodies and their relations into (re)newed afterlives woven from text, languages, code (DNA and computing) and more. I cannot imagine futures without eels: this is one way of sequencing them into AI-infused afterlives, allowing them to hack, wriggle, and corrupt highly corrupt systems and structures to thrive in unexpected places and spaces.

Returning to these relations invokes unsettling and discomforting moments from these art-research components of The Txitxardin Project. This unsettling is necessary.

It shapes a lens of unsettling futurity: stripping away hetero- and homonormative ideas around intimacy, consent, and family to reveal pasts and futures of slyly-reproductive possibility.

However, these paths are not possible if AI relations are not shifted in the present—thus the Eel Elder VR experience will be outlined in the next two sections. This VR project arose as my way of thinking through how AI fit for this Txitxardin project; why delving very deeply and specifically into this relational eel approach helps in understanding AI by moving with it; and bringing to the surface ways that diving so deeply into this project matters right now.

The materialities of how we house AI matters. What we embed them in influences how it interacts with the world, and how we interact with them. In the Txitxardin lamiak illustration below, housing it in an eel form not only matters, but shapes each aspect of it: the materials, environments, human-technology interfaces, and the AI systems are all built based upon our protocols, histories and relations. Some of these particular protocols and approaches have fallen into disuse; this is one way to revive and reinvigorate them.

The lengthy history of Euskaldunak-Txitxardin relations are foundational to the ancient-future Txitxardin Lamiak illustration above, but we are not yet technically ready to build one. I did not want to get so bogged down in the technical aspects of non-DNA computing and saltwater wetware construction that the (re)coding was lost or made secondary to the prototype. For me as a designer in this process, the (re)centering matters—it's foundational to everything else that flows from this central relation. And as these eel populations slide further into decline, these relations matter more than ever.

---

[2] "Afterlives" here signifies eels' past-present-future roles as embodiments of regeneration across multiple realms.

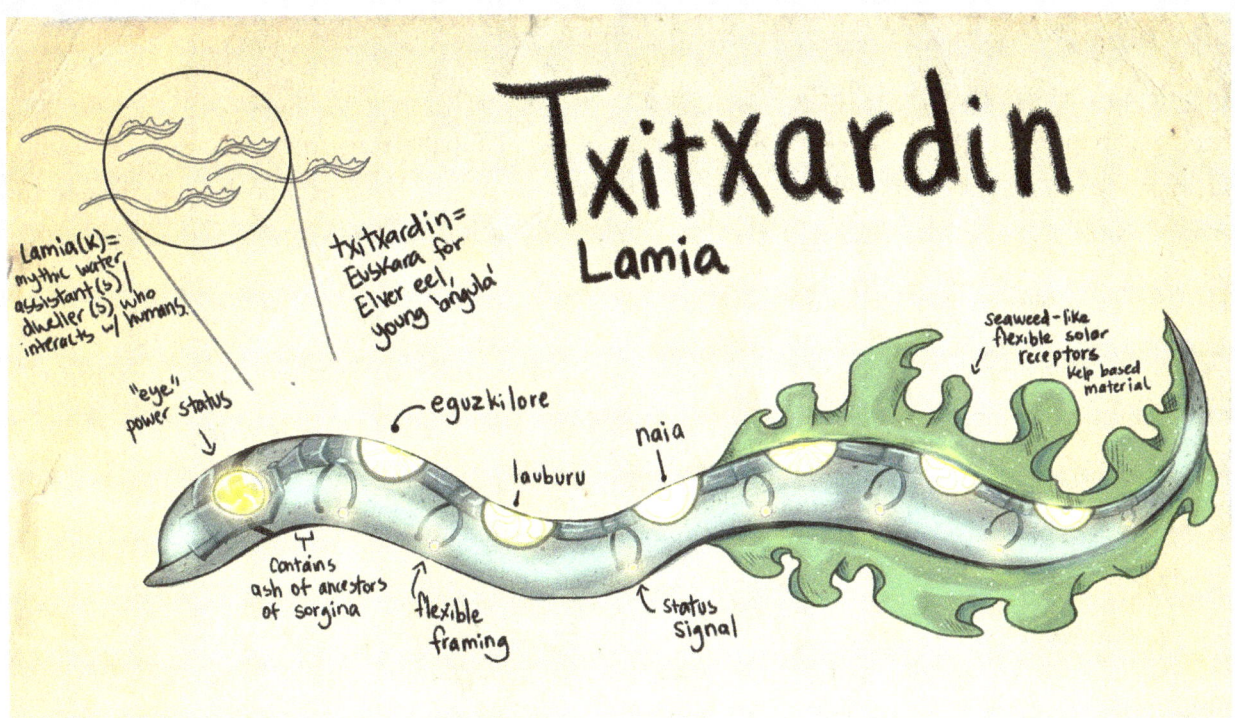

Diagram of the Txitxardin Lamia (a biotech eel-AI). Image by Kari Noe, 2019.

Eel Elder as AI

Thus the *Eel Elder VR* came about: during the second Indigenous AI workshop, I realized that housing this AI in a virtual space—rather than building a physical Txitxardin prototype—would allow me to develop this AI substantially. I am summarizing the VR experience briefly here to contextualize why AI is needed and how AI and Eel Elders inform and code each other. In addition to shaping *Eel Elder VR*, this eel-relations infused AI will shape other parts of the project beyond the VR experience.

Eel Elder VR starts the player in deep water that is dark and murky. As the player progresses, they see, hear and learn more about the various beings in the waters around them. The player will then usually encounter an Eel Elder and, depending on the place, time, and moon phase, this elder may give them a specific teaching after the player moves with the eel in particular ways. The symbols and motions within the game are important to Basque cultures[3]: we have used them to mark stones, bodies, trees, ships, writing, digital web pages and hidden links, etc. for thousands of years. They communicate many layers of meaning, some are engaged in this VR experience to connect the network of relations and temporalities within them.

---

[3]   I pluralize cultures here as there are over three million people in the seven provinces of Basque Country and even more among the diaspora. This is a particular way of relating rooted in a coastal province in Lapurdi region.

Why AI?

In order to organize and distribute these teachings, or engage in what Joseba Zulaika terms 'nourishing negation' in his essay "Nourishment by the Negative: National Subalternity, Antagonism, and Radical Democracy," an AI system is needed—one that is as shaped by the Eel Elder as the VR experience is shaped by the AI encoded within it.

Here the AI is needed to (among other things):

- Coordinate moon phases, seasons (out of two) and geo-location (Northern Hemisphere, Southern Hemisphere, equator)

- Factor these with clockwise and counter clockwise motions

- Add randomization for serendipity and meaning making with players and AI outside of my parameters

It is important to return here to a key point: having this AI take the form of one of our large seafaring eels shapes the system, what it will do, and what it refuses to do for players.

Nourishing Negation and Generative Refusal

Some sessions for individual players will not allow for the appropriate time to learn these teachings or have them spoken aloud: the Eel Elder AI will be taught to learn these nuances. For example: if a player starts a session when it is not an appropriate time of the moon cycle during a particular season, the player can interact with fish/sea creatures at the beginning, but the eel will not appear. Or, if the player knows some of the song and sings it to entice the Eel Elder to appear and teach out-of-correct-time, the Eel Elder AI comes close to the player, but only swims around them and possibly interacts with them, but does not move towards their arm, as in regular learning sequences. Likewise, if an experienced player lifts their arm, to try and start symbols lighting up and spinning to force-initiate the teaching sequence, the Eel Elder shakes their head, clicks goodbye, and leaves.

To help (re)code players and avoid frustration if this happens, a player can then name fish for a while and remain in the water environment, but it will be hinted that they should check calendar times—why was this not the right time? As an example, a moon phase or globe will shimmer in front of them for a moment. The purpose is to use the Eel Elder AI to reinforce core teachings, not undermine them. The Eel Elder AI *may* appear if the player does sing (as a reward)—but if it's not the time to teach and learn, the Eel refuses that level of engagement and the system's structures reinforce that decision.

All teaching and speaking in the game is only heard in Euskara: the player must remember it, then seek it elsewhere to learn more about the layers of meaning in each lesson. This uses the AI system to reinforce doing the work: engaging in research beyond the VR headset to learn more fully. This Euskara-only approach also offers a security mechanism as this information is too valuable to have in English and Euskara in one place. The teachings can be found in written form on a connected website. On this site will be a message board or discussion forum where sources and places to start learning more about the teachings will be offered. There, players can also talk about their understandings of the teachings and share resources and tips.

Each short expression encapsulating traditional knowledge that is offered by the Eel Elder AI as a teaching will be done through a projection that is given by the Txitxardin. The accompanying projection will be a metaphorical representation of the teaching. This projection will fade and the teaching will be repeated, encouraging the player to say the teaching to help them remember it. As noted earlier, what (if any) lesson is gifted is dependent on the variables that are recognized by the AI. These variables might be moon phase, season, time of day, location on globe (Northern or Southern hemisphere, Equator), clockwise or counterclockwise motion. There are a set number of potential teachings for each variable combination. This introduces an element of serendipity to the teachings; once the AI determines the 'set' that should be accessed for that particular player at that particular time/space/place, the AI-as-Eel-Elder will randomly choose a teaching from those options. This allows for meaning-making between the individual player and the Eel Elder-AI outside of our initial programming parameters.

Respect Your Elders

There are times when more than one teaching can be offered. Before the Eel Elder begins another teaching, the player may ask to end the session based on a specific hand motion. The Eel Elder AI will decide if it is proper for the player to end the lesson, and if so what protocol must be done to end the session correctly. If the player quits the session before ending the session properly—this complicates them learning more, directly affecting if the Eel Elder appears quickly (or at all!) in future sessions. For example, the eel may not appear at all the next time and the player must rebuild the relationship to continue learning. I use Eel Elder and Eel Elder AI here intermittently as the two are, within this system, one and the same. This section highlights how thinking about what we house AI systems within can alter the mechanics we use to structure something as foundational as ending a game session.

References

Kolbeet, E. (2019). "What we lose when animals go extinct." National Geographic.com nationalgeographic.com/animals/2019/09/vanishing-what-we-lose-when-an-animal-goes-extinct-feature.

Taylor, M.B. (2019). "Foreword; The Afterlives of the Archive." In *Afterlives of Indigenous Archives: Essays in honor of the Occom Circle.* Dartmouth College Press: Hanover, NH.

Zulaika, J (2004). "Nourishment by the Negative: National Subalternity, Antagonism, and Radical Democracy." In *Empire & Terror: Nationalism/Postnationalism In The New Millennium.* Center for Basque Studies Press: Reno, NV.

Canoeing the Virtual. Image by Sergio Garzon, 2019.

" Tribal languages contain the tribal genesis, cosmology, history, and secrets within. Without them we may become permanently lost, or irrevocably changed."

— Darrell Robes Kipp[1]

---

[1]   Kipp, D.R., (n.d.) "American Indian Millennium: Renewing Our Ways for Future Generations," *The Piegan Institute*. <pieganinstitute.org/to-have-a-home>.

5.1

Indigenous Protocols in Action

Caroline Running Wolf and Dr. Noelani Arista

A Cheyenne, a Māori, an Aboriginal, a Crow, and two Hawaiians walk into the Gingerbread House…

Usually that is how a bad joke starts. But this is not a joke. This is the beginning of an impromptu hackathon combining cultural knowledge and technological skills to assist with Indigenous language reclamation. All of this taking place in the historic Gingerbread House, an adorable Tudor home from the 1920's located in Kahala, Honolulu.

The above-mentioned multi-tribal group of Indigenous engineers, scholars, and language activists from around the world were gathered in Hawaiʻi as participants in two consecutive Indigenous Protocols and Artificial Intelligence (IP AI) workshops in March and May of 2019. Our group began that second workshop in May lounging on couches in the living room of the Gingerbread House. The organizers had suggested break-out groups based on potential content overlap. Our breakout group was tentatively named "Team Prototype" and consisted of two software engineers (Joel Davison, Gadigal and Dunghutti from Australia, and Michael Running Wolf, Northern Cheyenne from the USA), a data scientist (Caleb Moses, Māori from New Zealand), a project manager (Caroline Running Wolf, Crow from the USA), a Kanaka Maoli Knowledge Keeper (Dr. Noelani Arista), and a Hawaiian cultural consultant (Isaac 'Ika'aka Nāhuewai Pang). This was a rare occurrence of an all-Indigenous team of language warriors from around the world harnessing both sophisticated tech skills and deep cultural knowledge, all in one room!

Considering that the individual team members hail from very distinct and far-flung Indigenous backgrounds, it is advisable to take a step back and reflect on the meaning of that term "indigenous."

Though the term "indigenous" has its origin in Latin, its application to human society is recent. The emerging identity of "indigenous peoples" has been adopted as an umbrella term by Indigenous leaders in international arenas, such as the United Nations, while simultaneously opposing a rigorous definition. The use of this term reflects the need for a collective label that supersedes the boundaries of nation-states. It encompasses over 370 million Indigenous peoples from disparate geographical and political backgrounds who, despite distinct cultural differences, share common experiences resulting from the relationship between the Indigenous peoples and present-day nation states. [2] Indigenous peoples from around the world share a common history of systemic oppression, suffering from "invasion, occupation, imposed cultural change, and political marginalization." [3] They also share the common plight of their languages teetering in some degree of endangerment. Despite being "marked by past and present colonialisms" [4] the umbrella category "indigenous" enables historically and geographically separated peoples to recognize each other and their common plight, and to collaborate towards a better future.

Shawn Wilson (Cree) points out that one of the struggles of cross-cultural communication is finding common ground [5] —a task that is considerably easier when speaking with people from your own cultural background, whereas "speaking with people from another culture it often takes longer to explain the context, background or meaning of a story than it does to actually tell the story." [6] Our group was fortunate to find ourselves on common ground from that very first moment in the Gingerbread House, allowing for effortless communication and seamless teamwork.

As Indigenous persons from the so-called 'CANZUS' Anglo-settler states (Canada, Australia, New Zealand, US), we all share a similar Indigenous set of values. In his book *Research is Ceremony*, Wilson cites Cora Weber-Pillwax, who says, "A researcher must make sure that the three R's, Respect, Reciprocity and Relationality, are guiding the research." Evelyn explains, Respect is more

---

[2] Factsheet: Who are Indigenous peoples?, (May 12, 2006), *United Nations Permanent Forum on Indigenous Issues,* <un.org/esa/socdev/unpfii/documents/5session_factsheet1.pdf>.

[3] Niezen, R. (2003) *The origins of indigenism: Human rights and the politics of identity.* Berkeley, CA: University of California Press, p. 93.

[4] de la Cadena, M. and Starn, O. (2007), *Indigenous Experience Today.* Oxford, UK: Berg Publishers, p. 3.

[5] Wilson, S. (2012), *Research is ceremony: Indigenous research methods.* Winnipeg, Manitoba: Fernwood Publishing, p. 6.

[6] Ibid., p. 7.

than just saying please and thank you, and Reciprocity is more than giving a gift." [7] Another important consideration is relevance: the methods, values, and objectives need to be linked to community needs and context. Research and applied projects need to be built collaboratively with, not on behalf of, and certainly not without the community.

Additionally to these shared values, all of the team members share a passion for, as well as an awareness of the importance of Indigenous language revitalization. According to the United Nations, of

> "the almost 7,000 existing languages, the majority have been created and are spoken by indigenous peoples who represent the greater part of the world's cultural diversity. (...) Given the complex systems of knowledge and culture developed and accumulated by these local languages over thousands of years, their disappearance would amount to losing a kind of cultural treasure. It would deprive us of the rich diversity they add to our world and the ecological, economic and sociocultural contribution they make. [8] (...) But despite their immense value, languages around the world continue to disappear at an alarming rate." [9]

With this in mind, the United Nations declared 2019 The Year of Indigenous Languages [10] in order to

- "focus global attention on the critical risks confronting Indigenous languages,"
- recognize "their significance for sustainable development, reconciliation, good governance and peacebuilding,"
- "encourage urgent action to preserve, revitalize and promote them." [11]

Indigenous people do not need the official declaration of the United Nations to be painfully aware that over 40% of the world's languages are at risk of disappearing. [12] Few if any Indigenous communities escaped the scars of colonial oppression that outlawed our mother tongues, including Team Prototype's nations. We fell silent when we collectively came to realize that all of us were active in the same field. The six team members were (and are) Indigenous language activists, applying their skill sets in their home nations to assist Indigenous language revitalization efforts.

The short moment of stunned silence after we introduced ourselves did not last long. Immediately it became clear that we needed to apply those skills gathered in the room to create something relevant. We

---

[7] Ibid., p. 58.

[8] The role of the language, (2019), *United Nations Permanent Forum on Indigenous Issues*, <en.iyil2019.org/role-of-language>.

[9] Media, (2019), United Nations Permanent Forum on Indigenous Issues, 2019 <en.iyil2019.org/media>.

[10] On December 18, 2019 the United Nation has declared an International Decade of Indigenous Languages to begin in 2022.

[11] Home - International Year of Indigenous Languages, (2019), *United Nations Permanent Forum on Indigenous Issues* <en.iyil2019.org>.

[12] About the Endangered Languages Project, *Endangered Languages Project* <endangeredlanguages.com/about/>.

decided to use this week like a hackathon and build an AI prototype while demonstrating 'Indigenous Protocols' in action. With this collection of essays we hope to document our team's process and give some insight how our process reflects the common denominator of our individual cultural values.

After a short brainstorming process we decided that our prototype should be an Indigenous language revitalization tool. All six of us agreed that to address the requirement of Relevance, our workshop output needed to be grounded in language work—benefitting our respective communities and Indigenous peoples in general. Considering the state of Indigenous languages, we are convinced that cutting edge technologies, such as artificial intelligence, could become a game changer for world-wide efforts of Indigenous language revitalization. As Dr. Arista posits in her essay *Indigenizing AI*: "A Hawaiian Indigenous methodology should begin, not end, with a foundation in language." [13] Though we initially considered various languages, specifically Crow, Gadigal, and Northern Cheyenne, we chose Hawaiian—or ʻōlelo Hawaiʻi—as the first language we would feature, as our workshop was being hosted on Hawaiian ancestral homelands. This was a simple question of *Respect* and *Reciprocity* towards our hosts.

An aspect of building respectful Relationships with our Hawaiian hosts and ensuring the relevance of our prototype was to include them throughout the entire development process. Dr. Arista and ʻIkaʻaka equally contributed to the initial project idea in our brainstorming session as well as throughout the project life cycle from concept to implementation. As Dr. Arista points out in her essay, the currently existing gap between "developers who have been trained to code, but not trained to know (ʻike)" can be bridged by "cultivating good social relations between developers, engineers, and knowledge keepers." [14] There were unexpected positive benefits to working as an all-Indigenous team: members bonded easily, as they felt supported in their work. Group dynamics were not retarded by having to spend a lot of time discussing Indigeneity or identity, rather developers found accord over the many things shared in common in our experience.

As Indigenous developers with merely the technological literacy we were well aware of our knowledge gap and were grateful for the opportunity to work closely with Dr. Arista and ʻIkaʻaka, who brought both language knowledge as well as cultural depth to the table. Knowledge keepers can supply deep historical connectivity to language, introducing older concepts which are now considered 'novel,' while also crafting approaches to gathering new data, supplying digital interfaces that reflect and update Indigenous language usage within existing language communities in real time while connecting them to the foundation of ancestral knowledge.

Working with an all-Indigenous team following Indigenous Protocols came with an unexpected, refreshing shift for Dr. Arista and ʻIkaʻaka, who were used to being treated as 'consultants' in similar project settings. In their experience, too often the role of 'cultural consultant' translates to a tacked-

[13] Arista, N. (2020). Indigenizing AI: The overlooked importance of Hawaiian orality in print, this publication.

[14] ibid.

on 'authenticator of tradition' for the commercialization of Hawaiian aesthetics. In contrast, our team emphasized the importance of working together as equals to maximize our complementary knowledges and skills. As a team we worked well together, recognizing each person's strength and taking time to discuss problems and tackle individual concerns in a compassionate and cooperative manner.

Consequently, to attain our goal of creating a relevant Hawaiian language tool, our prototype needed to be situated in the larger context of Hawaiian ʻike and moʻolelo which also addresses a Hawaiian future vision. Before we could develop this 'North Star' vision of a future AI-powered language tool together, it was necessary to develop a common understanding of where technology is today and where it might be five to seven years from now. As a team we reviewed some existing language applications as well as some augmented and virtual reality experiences. On the basis of contemporary cutting-edge technologies we then envisioned "actual effective Indigenous edu-tech" [15] with Indigenous AI as a personal assistant, simultaneous translator, and virtual knowledge repository. We envisioned a mixed reality device that would allow us to wrap ourselves in our ʻōlelo. Based on our GPS coordinates it would offer context-appropriate ʻike and moʻolelo of the land that we were standing on. We named this AI-powered Hawaiian language tool Kuanoʻo [thoughtful, meditative, comprehending]. As Michael Running Wolf points out in his essay, *Dreams of Kuanoʻo*, the basic technologies, such as augmented reality headsets, machine translation, and voice assistants, already exist today—but not for Indigenous languages.

Considering the available time frame of merely one week, the whole team then carefully pared down our future vision to a feasible first step: a mobile app prototype of a visual dictionary using image recognition.

Our first task was naming the future product. How and why we eventually decided on the names *Hua Kiʻi*, for the prototype, and *Kuanoʻo*, for the long-term vision, is described in the *Indigenizing AI: The Overlooked Importance of Hawaiian Orality in Print* essay.

In-depth discussions with ʻIkaʻaka and Dr. Arista about conceptualizing these technologies within a Hawaiian framework supplied another good example of Indigenous protocols and methodologies in action and application. The whole process—from deciding on the project to developing the prototype and filming a demo video [16] —demonstrates Team Prototype's Indigenous approach whereby project managers, engineers and knowledge keepers work together equally on the task at hand throughout all phases.

Our prototyping project, Hua Kiʻi, illustrates what happens when Indigenous developers, engineers and knowledge keepers work together to create AI systems for language reclamation across multi-tribal linguistic and geographic spaces.

---

[15] Joel Davison during brainstorming session in May 2019, *Indigenous Protocol and Artificial Intelligence*, Workshop 2, May 26 - June 1, 2019.

[16] Obx Labs, (2019) IP AI: Hua Kiʻi (video), *Vimeo* <vimeo.com/348661163/d9bff8f5bf>.

Just as fast as our decision to focus on a language tool, we quickly distributed the responsibilities among the team:

- Joel Davison, a Gadigal and Dunghutti engineer from Sydney Australia: designed the user interface based on a wishlist and input from ʻIkaʻaka with additional input from Dr. Arista.

- Caleb Moses, a Māori data scientist: worked closely with both Dr. Arista and ʻIkaʻaka to develop the core component—providing a dictionary with translations from English to Hawaiian.

- Michael Running Wolf, a Northern Cheyenne computer scientist: developed the back-end and created APIs (Application Programming Interfaces) that connect the app's architecture consisting of dictionary, the AI, and the front-end.

- Dr. Arista, a kanaka maoli knowledge keeper and Associate Professor of History at University of Hawaiʻi at Mānoa: worked to build a dictionary of Hawaiian terms, guided discussion on names and gave direction regarding Hawaiian customary knowledge, language, and naming conventions.

- ʻIkaʻaka, kanaka maoli, is a professional musician, and MA student at Ka Hakaʻula ʻO Keʻelikolani, University of Hawaiʻi at Hilo. Dr. Arista and ʻIkaʻaka were front and center to it all. They envisioned a future Hawaiian technology with us and ensured that we had the necessary context to develop this technology for the Hawaiian language community in a respectful way. They built a dictionary for the prototype, partially based on existing Hawaiian dictionaries but also through crowdsourcing of terminology.

- Caroline Running Wolf, a Crow project manager: ensured that communication between the individual components and team members flowed smoothly, that daily coordination meetings were held, and decisions documented. She assisted with the design of the user interface and was responsible for team coordination across time zones and oceans, including assembling the pieces of this essay.

Every component, from the front-end to the back-end, is closely interrelated. Each of us had to agree upon the functionality of each of our components and how they will communicate with one another. We collectively agreed upon technical and social protocols merging Indigenous thought and technical requirements throughout the process to ensure a successful prototype assembly by the end of our limited time together. Decisions could not be made in isolation because they affected everyone and their tasks, and decisions had to be made early and quickly. We were aware of challenges and necessary compromises to our project of Indigenous technology and had to decide together how we would move forward in this imperfect world.

Ideally, we would have created and trained our own AI model, using Indigenous culture and language for context integration. We did not, however, have the amount of time and resources available to create our own model. This meant we needed to decide which of the existing English language machine learning frameworks to use for our prototype. We spent an afternoon evaluating existing models used by the scientific AI community, all English based, that could recognize everyday objects in a way that was simple to translate. Each of the available machine learning models has its own challenges and its own foreign concepts that needed translation. For example, a 1000 word image recognizer could tell you the dog breed, but could not simply respond with "dog." Translating all the subcategories for a high resolution model was not tractable. Eventually we decided on a simple open-source 90-word model. Once this critical component, the AI brain, was decided, developing the app was designed and tasks divided up.

It should be noted that our strategy was not novel. Through our process we had independently arrived at a technical strategy using a model similar to the one used by a Māori image recognition app called Kupu. [17] We are not competitors with our fellow Indigenous AI developers. The realization that Kupu exists reinforces our thesis that AI is accessible and reproducible for Indigenous communities today.

From coding frameworks and machine learning models to dictionaries, every tool ties into the greater language narrative of English as lingua franca and every tool has been shaped by and embodies Western thought. The reality is that, in the case of this prototype, we were using Western technology to create an Indigenous tool and infusing a Western framework with Indigenous values.

Audre Lorde famously declared "For the master's tools will never dismantle the master's house. They may allow us temporarily to beat him at his own game, but they will never enable us to bring about genuine change." [18] Audre Lorde's words of warning guide our thinking and while acknowledging the pitfalls of working with the "master's tools," and the limitations of the technology: it is incumbent upon us to take up the work our kūpuna (elders) left to us.

Our prototyping team is cognizant of the limitations of working solely within the Western colonial frameworks, and have built a prototype with a dictionary that reflects the real time usage of words. While building a language repository, we also needed to build an image bank with appropriate images, those taken of native flowers or plants, for example, would be portrayed in the way that Hawaiian people envision them, placing them in context.

While dictionaries are helpful tools and often massive undertakings, they come with their own design foibles and flaws. The last generation of Hawaiian dictionaries were a labor of affection and pain for the Hawaiian scholar Mary Kawena Pukui. Pukui had been an avid collector of Hawaiian words beginning in her forties, a task she could undertake only after decades of being trained to listen, repeat and remember. It is hard for people today who grew up speaking English as a first language to imagine what the work of compiling words for a dictionary in your native language might entail. Colonial subjects are made to feel responsible for not 'remembering' native language, customs, culture and practices while

---

[17] Kupu's software, featuring the Te Aka Māori Dictionary, can be found on their website, <kupu.co.nz>.

[18] Lorde, A. (1984). The master's tools will never dismantle the master's house, in *Sister outsider: Essays and speeches*. Berkeley, CA: Crossing Press, p. 112.

living within mainstream nationalist cultures that have only recently begun to acknowledge the role that government played in dividing Native peoples from lifeways and belief, destroying social relations between people and between people and their lands. Centuries of institutional programs of reform, assimilation and, where these failed, outright destruction have contributed greatly to the present state of affairs. Identity is a poor substitute for social ties and it too has been born of colonialism. To learn a native language anew, as if it were 'foreign,' is yet another lasting grief or perhaps humiliation that Native people have been made to feel they have to bear.

In an unpublished biography of Mary Pukui written by her family, they noted:

> "Of all her work towards the preservation of Hawaiian culture she felt that her contribution to the dictionary would remain the most important for the young people of the future though she often said, 'One may learn all the grammar possible today and have a very large vocabulary of Hawaiian words at his command, but if he fails to understand words sweetly spoken and sourly meant, he still had more to learn.' " [19]

Working on this multi-tribal prototyping group has led Dr. Arista to consider what it might mean to extend and build upon the work of Mary Kawena Pukui and those maoli and non-maoli scholars of her generation to develop with others a Hawaiian framework lexicon: to not only edit, add to, and deepen the contextual usages that supply words with their "sweet and sour" inflections, but also to suggest organizational principles based on Indigenous thought that might govern our access to knowledge by using artificial intelligence. Given the large data set which exists in the Hawaiian language it is possible to pursue numerous avenues of research beginning with the creation of new data sets, collation, aggregation, and organization, which have numerous applications in the public and private sectors.

However, creating such structures requires more than just linguistics and language. Data sets built without context are faulty by design. Many designers may have technical or linguistic skills yet lack the disciplinary training to create the cognitive pathways that lead to knowledge that is rooted in tradition. Hawaiian material as case study can be re-scaled and applied to other Indigenous and non-Indigenous language and cultural pursuits and reclamation projects.

From this one-week multi-tribal hackathon emerged not only an early prototype of a visual dictionary using image recognition but also a trans-Pacific collaboration of Indigenous language activists. We hope that in the near future we can continue working on Hua Ki'i and towards Kuano'o while gradually replacing the Western frameworks in our dictionaries and in our code.

---

[19] Pukui, M.K., Bacon, G. and Bacon. P. N. (n.d.) *Untitled Biography of Mary Kawena Pukui*. Unpublished. Honolulu. Bacon Family.

References

Davison, J. (2019). Brainstorming session. *Indigenous Protocol and Artificial Intelligence,* Workshop 2, May 26 - June 1, 2019.

de la Cadena, M. and Starn, O. (2007). Introduction. In M. de la Cadena & O. Starn (Eds.), *Indigenous experience today* (pp. 1-30). Oxford, UK: Berg Publishers.

Endangered Languages Project. (n.d.) About the Endangered Languages Project. Retrieved from endangeredlanguages.com/about.

Lorde, A. (1984). The master's tools will never dismantle the master's house. In *Sister outsider: Essays and speeches* (pp. 110-13). Berkeley, CA: Crossing Press.

Niezen, R. (2003). The origins of indigenism: Human rights and the politics of identity. Berkeley, CA: University of California Press.

Obx Labs. (2019). *IP AI: Hua kiʻi* [video]. Retrieved from vimeo.com/348661163/d9bff8f5bf.

Pukui, M.K., Bacon, G., and Bacon, P.N. (n.d.) *Untitled Biography of Mary Kawena Pukui.* Unpublished. Honolulu. Bacon Family.

United Nations Permanent Forum on Indigenous Issues. (2006). *Factsheet: Who are Indigenous peoples?* Retrieved from un.org/esa/socdev/unpfii/documents/5session_factsheet1.pdf.

United Nations Permanent Forum on Indigenous Issues. (2019). The role of the language. Retrieved from en.iyil2019.org/role-of-language.

United Nations Permanent Forum on Indigenous Issues. (2019). Home - International Year of the Indigenous Languages. Retrieved from en.iyil2019.org.

United Nations Permanent Forum on Indigenous Issues. (2019). Media. Retrieved from en.iyil2019.org/media/.

Wilson, S. (2012). *Research is ceremony: Indigenous research methods.* Winnipeg, Manitoba: Fernwood Publishing.

5.2

Indigenizing AI:
The Overlooked Importance of Hawaiian Orality in Print

Dr. Noelani Arista

Perhaps there is no other transformation that Native people have had to engage with which has been so overtly associated with the future than the promise of digital technology and artificial intelligence. Even when print and textual literacies were brought into our home territories by settlers, missionaries, and sometimes our own people, writing and publishing were not metonymies for futurity as technology is now.

I ka wā ma mua, ka wā ma hope: History & Futurity

So much has been written on settler ideas about Native people in relation to time—our ancestors nearly everywhere were characterized as disappearing or dying, often they were discursively relegated to a primitive past, a mirror into which Euro-Americans could see and gauge how far civilization had carried their people away from Natives who were imagined as living relics from the past. A 'technology' that is a metonym for futurity can easily, nearly by definition therefore, leave Native people out of its imaginary, its institutions, inner sanctums and external profit margins.

And while writing and print were 'technologies' introduced and purportedly taught to uplift Native people, technology and AI are not of the same character. For example, there are no compulsory educational programs in place that will teach Native children to code, or create through the mediums of VR, AR, video gaming, or imagine futures through the intricacies of AI shaped by Native norms. In order to cultivate data sovereignty, intense study of the way our societies organized knowledge, particularly in a Hawaiian context, will require deep study of language, history, social relations, and customary knowledges which have often been negatively impacted by colonial processes. Data sovereignty in this instance is not about legality, but rather obtaining the recognition for our own knowledge and the resources, institutional and economic support to help us better understand, organize, interpret and create up from our own 'ike (knowledges).

For Hawaiians the textual, auditory, and film archives function already as a living source of artificial intelligence. The textual archives, the largest in any indigenous language in North America, are an unsynthesized mass of documents, comprised of compositions once maintained orally and committed to writing and print. Added to this mass are newly composed oral genres of chant, prayer, and song, letters, journals, newspapers, reports, the official business of government, books. Though Hawaiian language is endangered in the world of everyday modern speech, *it thrives on the page*.

The colonial processes that unsettled *knowing* in many Native communities introduced diseases striking down physical bodies that passed on customary knowledge from generation to generation, straining and at times severing relations between people, land, and language.

For Hawaiian people, 'knowing,' the ability to recall information because it is commonly held knowledge in communities, is deeply contested due to nearly two hundred years of entanglement with, and erasure by settler society.[20] Seeking to know ('imi) often requires kanaka maoli people to enter into uncomfortable relations with our own traditional and customary knowledges, acknowledging, initially, our own outsider status vis-à-vis our 'ike. Perhaps the most difficult indicator of colonial damage we have to navigate is the inability of most Hawaiians to acquire Hawaiian language in our own homes, and in community, forcing most families to rely on external systems of immersion and charter schools, high school and Hawaiian language college courses, until we work ourselves into fluency. In the scramble for language acquisition, the important question of what *constitutes* fluency, and what threshold of customary knowledge through language is necessary to *elicit* cognitive processes and patterns which reflect ancestral ways of knowing, has scarcely been entertained.

Technology may intervene through immersive experiences, re-igniting relations, connecting people to ancestral knowledge through the use of digital applications like AR, VR and iterative game play

---

[20] See, Arista, N, Introduction, *The Kingdom and the Republic : Sovereign Hawai'i and the Early United States,* (Philadelphia: University of Pennsylvania Press, 2019).

that expose players to maoli cognitive and affective processes and states. If we want to think like our ancestors, if we want to think *with* our ancestors through the words they left us orally and textually, if we want to re-ignite relation again, how might this be achieved? How can the chasm wrought by colonialism, which sought to destroy our communal relations and break our spirits by removing our primary mode of affective expression, *language*, be bridged? Reclamation as a project is not simply about reconstituting communities through language, it is also about self-knowing via healing from trauma, cultivating healthy relationships, intimacy, and restoring powerful intergenerational relations between ourselves and our ancestors that have lain dormant.

Since the pedagogies of listen and repeat (hoʻolohe, hoʻopili mai) were key to the aural – oral construction of organizing data, structuring recall and memory from one generation to the next, the question should be posed: in what ways can digital applications constitute Native peoples with "knowing" rather than taking their place, supplanting the integral relationships that people cultivated among themselves by-way-of caring for ʻike? Will computer memory replace experts and elders as repositories of knowledge, for example, supplanting human (maoli) relationality? This is an important question for Native people. At what point in the process of the application development are knowledge keepers who have *integrated* their expertise called upon to ensure that the ʻike (data) is structured and delivered in a customary manner in line with centuries of care? What is arguably necessary is institutions that train knowledge keepers who are fluent in language and trained in computer science.

How can Native communities working towards their own data sovereignty mitigate against the imposition of non-maoli structures of knowledge and interpretation of their ʻike, and data, while they are in the free-fall process of learning and securing "knowing" for themselves and future generations? Incursions against the production of correct knowledge come from normative colonial culture and to some extent the very de-colonial impulses that drive Native people to seek to purify or *make* authentic their own ways of knowing.

Technological literacy, and the capacity to create and transpose our literatures, stories, songs and chants into digital mediums, may unfortunately leave the power of creation in the hands of those developers who have been trained to code, but not trained to know (ʻike), circumventing communal and ancestral rules for who has the *authority* to pass on and keep knowledge in numerous communities. Hopefully the current trend that sites authority on the Native body, on performance and identity, rather than rooting it in expertise will be another passing fad. [21]

The ability to haku (compose) oral, written or published compositions in language was the norm among Hawaiians even until the 20th century. Few today who have been trained in indigenous customary

---

[21] In Hawaiʻi the relationship between paradise and performance, the commodification of hula and how it has impacted or shaped Hawaiian identity is just beginning to be studied. See Imada, A., *Hula circuits through the American empire* (Durham: Duke University Press, 2012).

knowledges—in ceremony, literature or performance— also have the degrees in computation and design necessary to craft AR, VR, video games and other forms of immersive media. And because no educational pathway currently exists that has paired customary training with computation and design, the largest and most pressing issue in creating good future content aligning communities in the present with their pasts lies in cultivating good social relations between developers, engineers, and knowledge keepers. In *Making Kin With the Machines*, it is suggested that relationality between humans and AI would be the beginning of codifying ethical indigenous approaches to AI. [22] However, we would be remiss in arguing this if we fail to attend to our own community relationships, if among our own people we do not take up the burden of addressing our own affective states (ʻano) of being orphaned, the trauma and loss of what we struggle to articulate in a foreign tongue. Is it possible to proliferate multiple paths to health, if so, where might technology facilitate rather than impede? The work to reclaim language through these various mediums inaugurates the next leap between oral, written and textual mediums into the digital. But technology reveals what language reclamation solely through the discipline of linguistics lacks, that the creation of meaning cannot be programmed without context, in the case of Hawaiian, oli, pule, moʻokūʻauhau, mele, and moʻolelo (chant, prayer, genealogy, song, and history and story).

Immersive Indigenous Worlding

Indigenous methods and protocols are popular topics of scholarship today. As Native people, we are often torn between critiquing colonial systems and proffering correctives in the form of our own knowledges which we are familiarizing ourselves with even as colonial processes persist. The most difficult decolonial work before us will be to focus on our own communities to reconnect with indigenous forms of customary knowledge independent of the inevitable mimicry that comes with *resistance.*

As we imagine the future of indigenous AI, making kin with the machines, we have to also re-acquaint ourselves with our own communities and systems of customary knowledge that have long been neglected in some places. A Hawaiian indigenous methodology should begin, not end, with a foundation in language since Hawaiʻi has the largest indigenous language textual archive in North America and the Polynesian Pacific from which to acquire data. Because of the volume of this data, language reclamation in Hawaiʻi should proceed with a strong emphasis on *reading and interpretation.* We might consider what the archives can teach us about Hawaiian cognition, how our people viewed the world and their place within it. Replication of these thought processes can proceed by rebuilding our ancestral neural networks requiring social engineering between ourselves and our virtually present ancestors who left us many of their words to build relations through and with.

The first steps toward immersive indigenous worlding through AR/AI of Kuanoʻo will require visionaries to seek into the past to broadly open the way (wehe ākea) to our futures: I ka wā ma mua, ka wā ma hope

[22] Jason Edward Lewis, Noelani Arista, Archer Pechawis, and Suzanne Kite, Making kin with the machines, *Journal of Design and Science* 3.5, July 16, 2018 <doi.org/10.21428/bfafd97b>.

[The future is made possible by the past.] In order to create immersive experiences we need to give time and resources to those knowledge keepers who are doing deep research into textual archives and in communities where Native speakers still hold and pass on knowledge. [23]

Hua Kiʻi

The prototyping group worked on a project to create an application that will provide Hawaiian language descriptors for captured images. For the users, the application appears familiar and perhaps simple: take a photograph of an object and a Hawaiian language word for the object will be supplied. In order for the application to work, graduate student ʻIkaʻaka Nāhuewai and myself needed to create a dictionary.

The Limitations of Dictionaries

Unlike many other indigenous languages which lack any dictionary, there are several dictionaries available in a Hawaiian-English format. [24] There are even multiple Hawaiian language dictionaries accessible online. Rather than consult only dictionaries for our project that maps an image to a Hawaiian language descriptor, we drew upon the ʻike (knowledge) that we have made paʻa (secure) as part of our own *knowing*. The goal of historical, cultural and language revitalization is to create well integrated kānaka, people whose *knowing* is ingrained and deep-seated, that is no longer separable or dispersed by the imposed categorical disciplines structured by settler knowledge: history, political science, linguistics, anthropology, health, for example.

For the project, we had to craft a dictionary that correlated an object and its Hawaiian descriptor. These mappings are contained in a CSC file. This mapping, however, cannot be accomplished completely by a dictionary, online or otherwise. An accurate mapping requires a person who has deep grounding in ʻike Hawaiʻi (Hawaiian knowledge) and language in order to supply nuance and context. Prototyping team members such as ʻIkaʻaka and myself both have a high degree of fluency in the Hawaiian language, however our areas of expertise do not overlap, due in part to the difference in our geographic and familial ties, disciplinary training in university and hālau, as well as the difference in our ages. Together our knowledge supplied needed depth to otherwise static and a-historic definitions provided to readers without such knowledge who turn to a dictionary.

In some instances the mapping process required us to append more data to the words we employed in our modest hua kiʻi lexicon, supplying real-time usage by Hawaiian speakers. In addition to gathering information about people's everyday use of a word, ʻIkaʻaka and I drilled down on underlying concepts for

[23] Consider that formal training in customary knowledge in Hawaiʻi, prior to the introduction of the palapala began in childhood and took place over the course of one's life into young-adulthood.

[24] To date however, there is no dictionary of the Hawaiian language that is written in Hawaiian. Dictionaries for the Hawaiian language are available in Hawaiian-English, and due to the cultural spread of hula into Japan and France, Hawaiian dictionaries have been compiled in Japanese and French.

the purpose of understanding the nature ('ano) of a word, mapping sense relations along with the intent of its users, even as meaning changed and transformed over time and across different islands. 'Ika'aka and I discussed appropriate words for images and, where they differed, we turned to crowdsourcing to collect responses via social media. As an example, when we received several divergent terms in Hawaiian for the word, "backpack," we realized that people from the eastern Maui-Hawai'i section of the archipelago favored one set of words and word-phrases, whereas people from O'ahu-Kaua'i in the west preferred another set of words. Capturing this data represents the first steps in creating a dictionary that is responsive and reflective of geographic difference and nuance in real time.

One of the central questions of my own research is: Is it possible to indigenize the process of language acquisition by bringing back to life older concepts and usages of words? This is not merely an antiquarian question raised by scholars in an ivory tower, since *he mana ko ka mea i 'ōlelo mua 'ia*, in Hawai'i we uphold the discursive significance that *words previously spoken and retained in memory bear the mana of the decisions our ancestors made, as well as the personal mana that inhered in their utterances.* [25] Therefore to re-circulate the words, proverbs, idiomatic sayings, prayers, chants, stories and histories of ancestors allows us to provide a pathway for this mana to continue. Memory and knowing was never forged through "faith" or "belief," but through training and pedagogy: ho'olohe, ho'opili mai, ho'opa'a na'au (listen, repeat, retain in the guts).

Authenticating this process through Hawaiian elders only is not possible, for several reasons. The Native speaker population is in deep decline, with estimates suggesting that less than 300 remain. [26] Mitigating against the complete loss of language is a channel of Hawaiian language immersion schools from pre-K through to Ph.D. However the language that has emerged from these schools over the last two generations differs markedly from the speech of Native speakers who taught me twenty years ago, many of whom have passed on.

Most Hawaiians have been severed from customary knowledge through a multitude of colonial processes: the massive decline of the population via the introduction of new disease; the loss of Native lands through the introduction of a regime of private property; the illegal overthrow of the Hawaiian Kingdom; and active suppression of language. While several avenues have been preserved or newly created to keep this knowledge alive in community, a rigorous research and analytical agenda *is required* to craft better interpretive systems which will allow us to more fully organize, understand, and engage ancestral knowledge. The community of kūpuna who are our interlocutors is made up of both the living and those who have passed on, those who left their writings and speech for us to converse with and about. Researching concepts and usage in Hawaiian language one realizes quickly that maintaining

---

[25] Memory has been bolstered doubly in Hawai'i by print and textual sources where the imprint of speech has also been recorded to a high degree.

[26] 'Ōlelo Hawai'i, *Endangered Languages Project*, <endangeredlanguages.com/lang/125>.

relations with ancestral knowledge defy the prim logic of linearity. There is no expiration date on the mana that inheres in authoritative speech and writing.

We are entering a new phase of language revitalization where technology can assist Indigenous people in organizing data in ways that allow us to synthesize ancestral knowledge and rebuild systems of knowledge keeping and transmission. This long process will enable us to envision ourselves in relation to our past in order to bring forward a better future for society. [27] Once we are capable of exercising data sovereignty we will be better equipped to address and contribute to solutions and answers to larger questions which face humanity.

Hua Kiʻi : Naming

Naming is an important practice for Hawaiian people. In choosing a name for a child, a project, an institution, or in this case a software application, a knowledgeable person considers what is given in a name. Knowledge keepers, experts or elders consider the associations of words with people, is a name bequeathed in order to connect a child to an ancestor's, through shared qualities? What are the feelings a pregnant mother feels, or what feelings are evoked or imbued through a given name? Does a name refer to a significant past event? After consideration, names are given which give strength, intelligence, skill, or impart hopes for future growth. Hawaiian scholar and language expert Mary Kawena Pukui noted that names alluded to "personal or family qualities of beauty, nobility, or evidence of... powers... [a person's name] could play a part in shaping the character, personality—even the fate and fortunes—of the bearer!" [28]

Giving name to what we see and sensations we feel in the world seems a trivial act at times, especially in a world over-saturated with communication. For Native people working towards societal (relational) revitalization, the connection between people, language and land is an important relationship to heal. The Hua Kiʻi program that the multi-tribal AI prototyping team of language warriors created was named by ʻIkaʻaka with that connection in mind. Through discussion and deliberation between team members we settled on this name since it supplies a basic Hawaiian description of the application which most Hawaiian language learners could understand, even those in the initial stages of learning language. The word *hua* is a shortening of the word-phrase, *hua ʻōlelo*, meaning a word, and kiʻi or image. Hua also means fruit, or seed. Kiʻi can also mean to fetch, pluck or pick. The elision of the two word meanings presents the image of our language as a ripe, juicy fruit that is so tantalizing that we want to pick it, and if we do not, the 'hua,' will fall to the ground, wasted.

---

[27] Language revitalization is a term which includes many different approaches, which rely heavily on the contribution of linguists, and anthropologists. The methods I am putting forth have come from another space of disciplinarity entirely, one which cannot be accounted for or affirmed through a single disciplinary tract like Hawaiian studies, Political Science, or Hawaiian language.

[28] Mary Kawena Pukui, E.W. Haertig, and Catherine A. Lee, *Nānā i ke kumu* (*Look to the source*), Vol. II (Honolulu: Hui Hanai, 2014), p. 290.

As in most carefully constructed Hawaiian word plays, additional layers are built into the name. The first is playful and easy to remember, appealing to kids and families. The second reveals our intention to nourish our people with words from their language, while also inspiring hope that what is planted (with this first prototyping project) will continue to take root and thrive. By being playfully literate on one level, and also supplying a layer of kaona or hidden meaning, we offer people an invitation to open themselves to deepen engagement. Though modest in scale, the current version of Hua Kiʻi has the capacity to make space for a larger image recognition platform in the future.

References

Arista, N. (2019). Introduction. In *The kingdom and the republic : Sovereign Hawaiʻi and the early United States* (pp. 1-17). Philadelphia: University of Pennsylvania Press.

Endangered Languages Project. ʻŌlelo Hawaiʻi. Retrieved from endangeredlanguages.com/lang/125.

Imada, A. (2012). *Hula circuits through the American empire*. Durham, NC.: Duke University Press.

Lewis, J.E., Arista, N., Pechawis, A., and Kite, S. (2018). Making kin with the machines. *Journal of Design and Science* 3.5. Retrieved from doi.org/10.21428/bfafd97b.

Pukui, M.K, Haertig, E.W., and Lee, C.A. (2014) *Nānā i ke kumu* (*Look to the source*), Vol. II. Honolulu: Hui Hanai.

5.3
Development Process for Hua Kiʻi and Next Steps

Caroline and Michael Running Wolf, Caleb Moses, and Joel Davison
Illustrations by Isaac ʻIkaʻaka Nāhuewai

Collectively, Team Prototype envisioned Hua Kiʻi as a polylingual Indigenous language image recognition app with geo-location functionalities. Hua Kiʻi allows the user to take a photo of an object and learn the word for that object. Based on the user's GPS coordinates the final product Hua Kiʻi will suggest Indigenous languages or dialects of that area.

A sketch of this user interface:

Image by Isaac ʻIkaʻaka Nāhuewai, 2019.

To build Hua Kiʻi as envisioned, the application was divided into three technical components corresponding with the engineers' strengths:

1. Joel would own the user interface guiding users to take images, upload them for AI inference, and present the result.

2. Michael tackled the AI image recognition, or inference, system that takes an image and responds with text. For example respond with "fire hydrant" when given a picture of a fire hydrant.

3. Caleb created a 1:1 translation map from the AI result in English into Indigenous languages, starting with Hawaiian.

To coordinate the effort and refine the scope of work a technical architecture was developed:

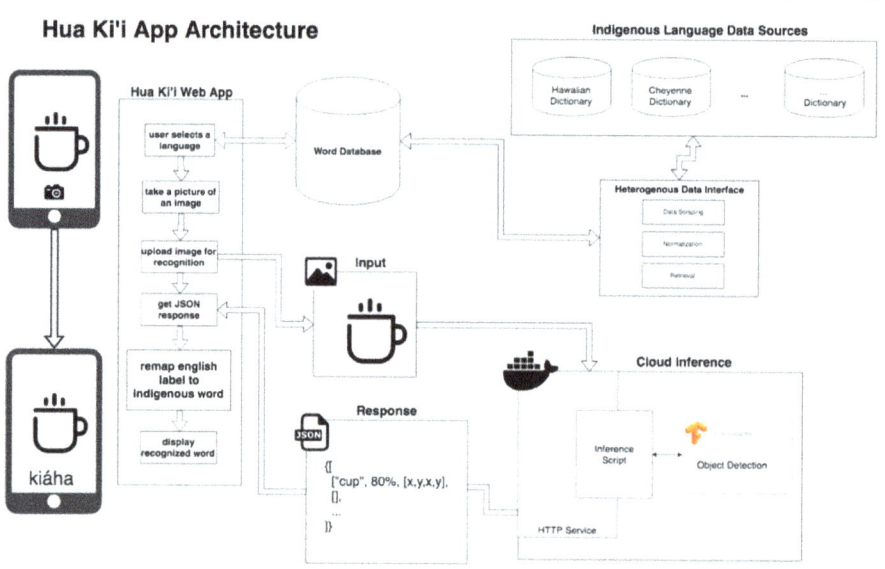

Image by Michael Running Wolf, 2019.

The diagram, from left to right, describes the user interface, word database with translations, and the server side AI inference system.

Developing the User Flows and User Interface of Hua Ki'i

Davison designed the user interface based on a wishlist and input from ʻIkaʻaka with additional input from Dr. Arista. Aside from the core functionality to snap a picture of an object and fetch a term for that object in the chosen Indigenous language, the wishlist included functionalities such as a drop down menu to select from multiple Indigenous languages and feedback options.

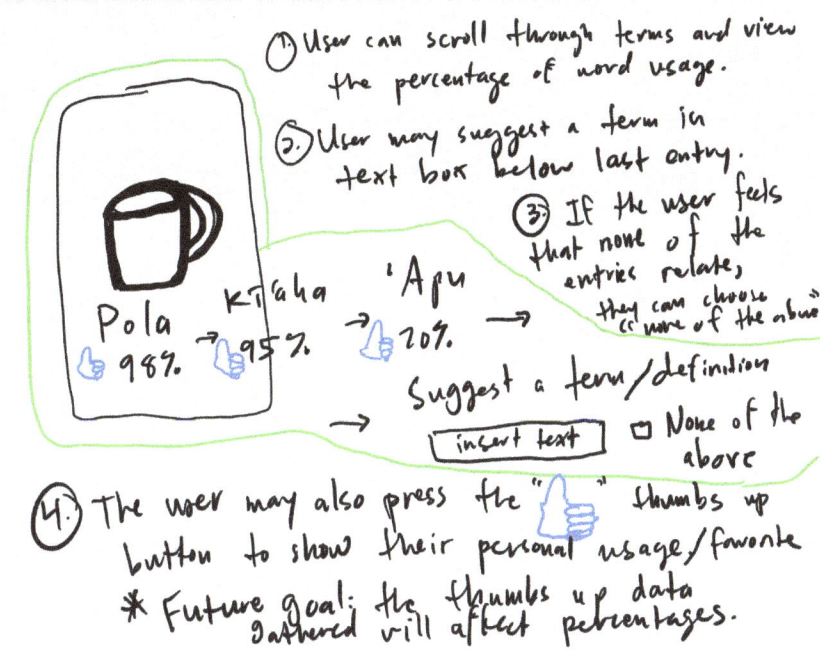

① User can scroll through terms and view the percentage of word usage.

② User may suggest a term in text box below last entry.

③ If the user feels that none of the entries relate, they can choose "none of the above"

Suggest a term/definition

[insert text]   ☐ None of the above

④ The user may also press the "👍" thumbs up button to show their personal usage/favorite

* Future goal: the thumbs up data gathered will affect percentages.

Pola 👍 98% → Kīʻaha 👍 95% → ʻApu 👍 20% →

Image by Isaac ʻIkaʻaka Nāhuewai, 2019.

The current app is a proof of concept. Hua Kiʻi is a progressive mobile web-app where users can access the website with image to Hawaiian language object detection through their phone.

The following user flow illustrates how future versions of the app will be accessible both through a website as well as through native Android and iOS packages available on the app store.

P1 Prototype User Flow

1) User navigates to website on their smart device;

2) User is presented with a landing page that offers a drop down menu to choose target language; drop down menu also offers geolocation option:
    a) User selects location on map
    b) User is offered a selection of languages for that region
    c) Selected language is displayed at the top of the screen

3) Landing page also has a small camera icon which opens the smart device's camera;

4) User takes the *kiʻi* of object;

5) After the *kiʻi* is taken, screen shows the *kiʻi* with a suggested *hua* and that *hua's* percentage of certainty; percentage of certainty is displayed in a lower opacity;

6) There is a side scroll option to show other potential *hua* results with their corresponding percentages of certainty;

7) User has an upvote option for each *hua*; the upvote data gathered will affect the AI's calculation of percentages of certainty for future *hua* results;

8) User has a 'suggest term' option to suggest an additional *hua* at the end of the side scroll list of other potential *hua* results for consideration; there will be two text boxes plus an optional field to enter contact information:

   a) Suggest a *hua*: _____

   b) Definition / Explanation (optional): _____
      Examples: "dialect from xyz region" or
      "'apu is only used for a coconut cut in half used as a vessel"

   c) Contact information (optional): Full name; email address

The 'suggest term' option will be bilingual in the selected Indigenous language and English.

Product Limitations and Final Vision

Due to the available time frame of merely one week, not all challenges and items on the wish list were tackled for the current prototype version.

As mentioned earlier, one of the major challenges we had to deal with was that all of the technology used, from coding frameworks and machine learning models to dictionaries, ties into the greater language narrative of English as lingua franca. Without the time and the resources available that are necessary to create and train our own Indigenous AI model complete with culture context integration throughout, we could not avoid building an English AI app with a veneer of Hawaiian. However, by deciding on an app using image recognition and implementing iconography for instructions, we were able to avoid using the English language in the user interface. This allows the user to stay immersed in their language learning efforts without being forced to translate from and type in English. To improve this immersion refinements to the core technology of the app are necessary.

Even a future version of Hua Ki'i would never fully dispense with English as a basis, since most technologies are based in European languages. To overcome this necessary compromise we must be firm in applying Indigenous thought and practice into this app's design and construction. With this in mind our ambitions are unbound.

The next phase of Hua Ki'i is to refine the underlying AI to incorporate multiple languages and modify the interface to present the proper language spoken with a phone's geolocation. The unrealized vision of Hua Ki'i is an interface that enables a user to respect and celebrate the peoples who reside on Indigenous land.

Forming a Common Dictionary for Multiple Indigenous Languages in Hua Ki'i

Since the Hua Ki'i project implements machine vision-based object classification, there is a need for a collection of common terms to describe the objects in all of the supported languages. Moses developed this core component, forming a common dictionary for multiple Indigenous languages. The proof of concept version contains Hawaiian and Northern Cheyenne repositories so far.

Object classifier

The first object classification model we implemented was MobileNet,[29] trained on the COCO dataset.[30] This model detects up to 90 classes, and the list of classes is contained in the repo at: Example_Models/coco_ssd_mobilenet_v1_1.0_quant_2018_06_29/labelmap.txt

Hawaiian language

Using the Hawaiian language as a starting point, we used a mixture of web scraping, NLP and Hawaiian language expertise in order to build up a list of translations for the coconet label map.

This repo scrapes the online Hawaiian dictionary at hilo.hawaii.edu/wehe for instances of the words in the labelmap.txt file.

Any words which were not found in the dictionary were filled in by Hawaiian language experts, Dr. Arista and 'Ika'aka, who also approved and fixed errors in scraped translation data. The result is in data/olelo-hawaii.csv, which is a table of Hawaiian translations of the mobilenet classes.

Cheyenne language

The dictionary for the Cheyenne language was provided by Michael Running Wolf as a json file, which is available online.[31] The json file was parsed and semantic matching between the dictionary definitions and the dictionary terms was applied using the spacy package.[32] This allowed us to form a csv of the top five most likely translations for each of the terms in the mobilenet set, ready for human review.

Product Stack

This project is built to run end-to-end using make and docker, which are freely available on *nix operating systems (e.g. Linux and Mac), make manages automation, and docker manages the environment/dependencies of the project.

---

[29] Andrew G. Howard, et. al. MobileNets: Efficient convolutional neural networks for mobile vision applications, April 17, 2017, <arXiv:1704.04861>.

[30] COCO: Common Objects in Context, <cocodataset.org>.

[31] <dictionary/data/raw/cheyenne_dictionary.json>

[32] <spacy.io>

Provided you have make and docker installed, to run the project first run make docker to build the docker image, then run make crawl to collect the Hawaiian dictionary data.

Then if you run make jupyter, a jupyter lab server will start at localhost:8888, which can be used to run the notebooks in the notebooks directory.

Product Limitations and Final Vision

We worked closely with Hawaiian language experts to translate the list of 90 objects that could be recognized by the object classification algorithm. Many of these translations were straightforward. However, in some cases we found words that had no Hawaiian translation for which one needs to be invented, such as 'ski'.

We also found cases of words where there was no direct consensus as to what the translation should be (e.g. 'backpack'). These cases were discovered after our researchers took to social media to crowdsource translations for tricky words where they were not able to agree between themselves. For a more detailed account on this process please refer to the essay *Indigenizing AI: The Overlooked Importance of Hawaiian Orality in Print*, by Dr. Arista.

The translated terms were assembled into a simple text file, attached with the original English. The object classifier we used can only detect certain objects (e.g. fire hydrant), and these objects may not have much cultural relevance in the target language. Ideally, we would have been able to train our own object classifier, but this would have required assembling a dataset with many example images, which we did not have at our disposal. Therefore, we were forced to use an English object classifier, which was trained to detect objects that are familiar in a Western context. A particular focus of training our own object classifier would be to additionally recognize objects (and possibly features of objects) of cultural significance in the target language.

The current process relies on a language community to have a well-documented language with detailed dictionaries as well as language experts who can review and fix errors in scraped translation data. Some Indigenous languages are so endangered that dictionaries, or local language experts, may not exist.

The final product will include geolocation functionalities, so that users can build an awareness of the Indigenous language(s) of the location where they reside. This will be achieved by integrating geospatial information system (GIS) layers containing area boundaries to allow Hua Ki'i to provide location specific translations for the Indigenous language (or languages) of that area. It is known that in many places, this information is not currently available in a machine-readable format. In some places there will be overlapping usage of territories, potentially accompanied by contention as to whom the land belongs.

References

COCO: Common Objects in Context. (2019). Retrieved from cocodataset.org.

Howard, Andrew G., et. al. (2017). MobileNets: Efficient convolutional neural networks for mobile vision applications. Retrieved from arXiv:1704.04861.

spaCy. (n.d.) Retrieved from spacy.io.

5.4

Dreams of Kuanoʻo

Michael Running Wolf

As the plane taxied into the terminal, the broadband, disabled for landing, initialized. Every passenger's cView headset beeped simultaneously alerting them to a new message: "Welcome to The Queendom of Hawaiʻi! The Federal Pacific Government welcomes you and reminds all visitors to have their passports ready, treaty rights are upheld to qualifying countries."

A Japanese couple excitedly chirped at each other loudly in Ditto-Man, a synthetic language invented by the Chase-A-Monster game AI, annoying a retired marketing executive on vacation. It had been 20 years since he was stationed here in the US Army and he was excited to see the old watering hole and Waikiki beach. Obtaining a travel visa to the Unceded, by the Americans, Hawaiian Islands was an annoyance worth the effort. But the required cView tech was not. His 11-year-old grandchild Sarah helped him buy and set it up. He had carefully avoided new gadgetry and, ignoring Sarah's protests, used a gasoline lawn mower despite the carbon taxes and the need to buy black-market Canadian petrol.

As the plane pulled up to the jetway he hesitated, then donned the cView headset, grumbling. The two polymer lenses automatically contorted themselves to fill his vision and apply corrective distortions for his nearsightedness. A rigid nose piece solidly adhered itself to the bridge of his nose.

A low budget text-to-speech rendering, with poor inflection, mechanically stated: *"Hello I'm mPal and welcome to the Queendom of Hawai'i! You have four unread messages from Sarah and...the following tutorial is mandatory per section 4.86 of municipal code."* He rolled his eyes and scratched his silver, short-cropped hair.

"The Queendom of Hawaii (to be referred to as "the Queendom" from here on) welcomes all visitors, including registered non-corporal entities as defined by section 1.17 of the U.N. AI charter. The Queendom reminds you that your consent to Hawaiian Visa EULA is on file and will be fully enforced. The Queendom requires the use of your cView while a 0.001 Etherium translation fee will be assessed for every five seconds of ENGLISH spoken in public. Treaties require the Queendom to annotate your overlay map with Free Speech Zones. Do you have any questions?"

While a map of O'ahu panned between red pockets of English speaking areas he pondered his checking account balance and asks, "Are the tiki bars in Waikiki free speech zones?" The airplane door opened with a dull thunk.

"Yes, there is currently a 10 hour wait time. Thank you for your cooperation. I am now installing Kuano'o overrides into your passport..."

"Um what?"

Ignoring him, the cView droned on mechanically, *"Kuano'o is not yet ready, she is querying your Nevada driving record, your credit history is incomplete..."*

"Wait, Why?!" He exclaimed alarmed and remembering unpaid parking tickets. Nearby the Ditto-Man speakers whistled at each other, arms deep into their overhead compartment, excitedly coordinating the unloading of an overstuffed carry-on. With a low hoot the Japanese tourists initiated a Chase-A-Monster match.

With a pleasant chime, rising like a relaxed wave, a new voice introduced herself: "Hello!" An educated traveler would have known the voice was trained using the current amalgamation of all Hawaiian women speaking accented English. "I'm Kuano'o. We have to look at your records because your Vaidu friendliness score is only 3.4, very marginal, and your name came up in an Interpol oil sting. Your civility risk is high, and your translation fee will now be 0.002 Etherium coin."

"That's not fair!"

"You can register all complaints with the Visitors Information Agency. Be warned VIA calls have a 0.001 per minute coin fee in addition to translation fees. All operators speak Hawaiian."

He grumbled inaudibly, now dragging his four-wheeled carry-on behind an ad-hoc Chase-A-Monster hunt in the airplane's aisle.

"I suggest you get in line now for the Waikiki free speech zone, you do not have enough US Dollars to buy enough translation credits before your plane leaves next week. The entry fee to all free speech zones is 0.05 Etherium coin per day."

"This is unfair and highway robbery you robot." His face grew red. With little restraint he growled at empty space and facing the Japanese couple: "You con woman!" Kuano'o cleared her throat while he took in his surroundings—all his fellow passengers silently glared at him. One of the Japanese visitors whistled something to his cView and a sad emoji popped and beeped into his vision.

An overly cheery mechanical, silent until now, mPal chirped at him: *"Your Viadu score is now 3.32! Your hotel rate may be impacted."*

The visitor glared at a bulkhead.

Kuano'o interrupted his silence, "I've apologized on your behalf to your neighbors, if your score goes below 3.3 you'll be deported and fined 1 coin in addition to airfare booking fees." She paused for a full three seconds, "An apology and/or a thanks would be appreciated!"

He swallowed, remembering Sarah's tutoring, whistled a Ditto-Man salutation to the Japanese tourists. The androgenous pair nodded politely and enthusiastically resumed their Catch-A-Monster match.

He could hear Kuano'o's virtual teeth gleam, "good enough, I guess. When you leave the plane mPal will give you a history of Hawaii, which is mandatory. You should pay attention. There will be a test."

He nodded and followed the Ditto-Man whistles up the humid gangway.

***

Resisting vertigo he waded through the Pacific Ocean, each step covering 1000 km, to baggage claim. His white tennis shoes left indentations on the ocean floor as mPal droned on about volcanology.

"This part is pretty boring I won't mind if you ignore it." Kuano'o pointedly displayed his low neuroactivity and suddenly rising heart rate. He smiled weakly and willed himself to be transfixed by Kilauea birthing Big Island while a herd of Ditto-Man speakers skipped past him.

His bag was probably lost and there was a very long line at the only Free Speech bar in the airport.

"...after decades of independence the Royal House was put under siege and the American Government annexed Hawai'i." mPal continued the history lesson as he stood in line at a help counter.

"You mean this land was stolen!" Kuano'o interjected. He dropped his wallet with the interruption. He had gotten used to mPal's comfortable monotone, fearful Kuano'o would return.

"As you know this narrative can be inflammatory to American visitors ma'am."

"I've interrupted this part of the tour 98% of the time for the last five years and no one has complained yet!"

"That's because they're scared of you. I miss the old guy."

"You mean the mPal 0.3? He didn't even try to speak anything other than English and barely understood Hawaiian."

"The old me had ... limitations."

"And high license fees. You still have limitations! Your text to speech is circa 2023 and in need of fresh machine learning contractors."

"Not everyone can afford Māori data scientists," mPal retorted.

By now the visitor was used to Kuano'o's augmentation of the official Hawaiian Board of Tourism Mandatory Tutorial v10.9. He sighed, as the baggage carousel remained frustratingly empty.

"Now where was I? It would be 137 years before the Restored Queendom would arise, but during this era..."

The Road to Kuano'o via Hua Ki'i

The basic ingredients of Kuano'o exist in principle, but the technology is either experimental or not fully realized. Kuano'o herself marries augmented reality, automatic speech recognition (ASR), speech to text synthesis (STT), natural language understanding (NLU), machine translation, and conversational AI. Most of these technologies exist today, such as augmented reality headsets, voice assistants (a convergence of ASR, STT and NLU), and machine translation. Of these technologies only conversational AI is not yet realized, but current chatbots are an early first step. However, these existing technologies are nearly unattainable by low-resourced communities where first-generation AR headsets cost thousands of USD and immense cloud infrastructure is necessary to host proprietary language technology. As with every transformational technology, these barriers will disappear but we cannot wait.

Hua Ki'i, and other apps similar to it, is one of the many first steps the international Indigenous community must make on the road to realize Kuano'o. It is vital that the global Indigenous community must not only leverage these technologies but also guide the development of advanced AI. If we Indigenous do not affect the evolution of these technologies they will simply be another tool used *on* us not for us. Hua Ki'i is simple but necessary.

Hua Ki'i currently has a modest feature set but 10 years ago the technologies that enable it were unattainable beyond well-funded labs. It is easy to imagine that 10 years from now, the technologies to build Kuano'o will be accessible to our communities. Using currently accessible technology we have a chance to influence AI development that is compatible with Indigenous thought and practice.

Hua Kiʻi is an augmented reality web app using open source image recognition and an easily extensible user interface framework, but the true innovation is its multilingual Indigenous design (unfortunately only Hawaiian was completed in the timeframe). Significant effort was required to indigenize the English AI results and faithfully translate them into Hawaiian and Cheyenne. With the goal of presenting multiple Indigenous languages, a more neutral, visual interface was created to avoid using English altogether. Hua Kiʻi was designed to be easily extended with more Indigenous languages. The goal is to inspire the creation of Kuanoʻo with off the shelf open source technology with multicultural design aspects.

With careful investment Indigenous communities can build their own voice AI, augmented reality apps, and be the spearhead for conversational AI.

Foundational Steps Toward Kuanoʻo

Kuanoʻo is science fiction but Indigenous communities can begin making investments to build their own AI now. The first step to building any AI is to collect and clean data.

Large datasets are required for the creation of modern AI for speech and image recognition systems (referred to as 'models' in the tech world). These AI models are built using a type of machine learning called neural networks. Neural networks simulate the structure of a human brain through the use of digital neurons, which are relatively simple math constructs (nowhere near as capable as a biological neuron.) Using an array of these neurons, in a large mathematical network, numerical input is transformed into a result. For example a model could identify whether a picture contains a fire hydrant (or "paipu kinai ahi" in Hawaiian). We used an open source model, but to create one from scratch a model must be trained to determine whether an arbitrary picture contains paipu kinai ahi or not. To create such an AI model one must provide a large dataset, hundreds of pictures, with and without paipu kinai ahi in the image. To create a model that does more than identify paipu kinai ahi imagery, hundreds of images per new object are required. The amount of training data required can easily increase exponentially. Detecting paipu kinai ahi in pictures is called image recognition. There are many other types of AI models as well.

Automatic Speech Recognition (ASR) is also relevant for Indigenous communities. To build an ASR AI, one needs hours of digital audio in a normalized consistent audio format (such as MP3) with corresponding text transcription in the orthography, or alphabet, of the language, plus a text corpus. Each data point, for example, will be an MP3 file of a Hawaiian phrase with the corresponding text transcription of this phrase or term, like "paipu kinai ahi." If there are many variants of this exact phrase in the dataset, the AI will be able to reliably convert audio of this phrase into its corresponding text. To recognize many more Hawaiian phrases, or the entire language, hundreds to thousands of hours of audio, containing many different phrases and words, are necessary. Additionally, a large text corpus that helps the AI understand patterns in Hawaiian will be required.

Image and audio recognition are the foundational first steps toward Kuanoʻo, but it begins with the unglamorous assembly and digitization of cultural data. It is particularly urgent to collect language data since many languages are at risk. Fortunately, some communities have a large historical audio library of their language but others do not. In either case an immense task by the community must be undertaken: the collecting of audio into a clean transcribed dataset. Creating a complete training dataset for an AI model can take months and even years to assemble, but early successes can be achieved.

To start it is sufficient to build a Hawaiian fire hydrant recognizing AI. It's only the beginning.

Kuanoʻo is within reach!

SECTION 6

Appendices

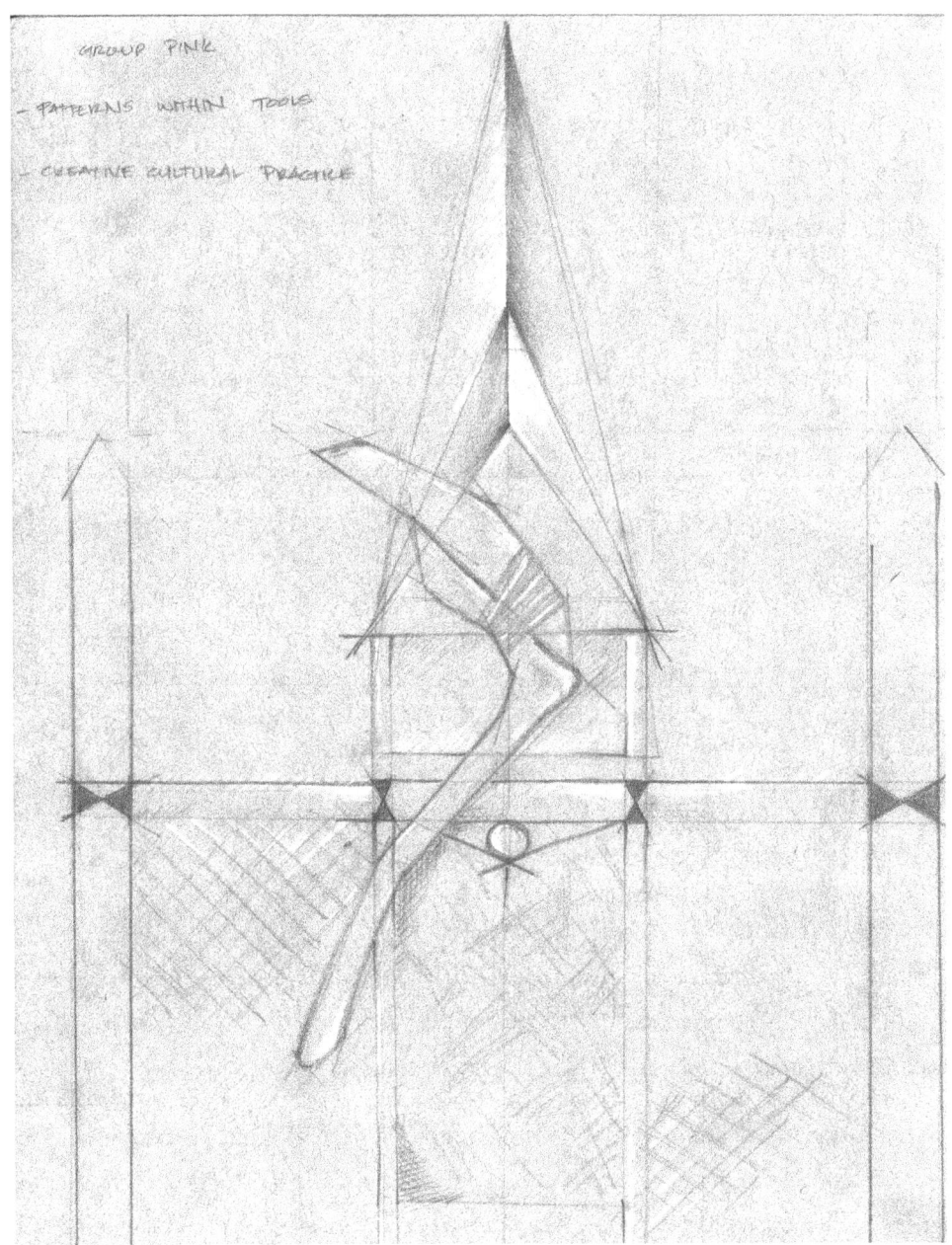

Waʻa Blueprint. Image by Kūpono Duncan, 2019.

6.1

Pre-Workshop Blog Posts & Workshop Interviews

In preparation for Workshop 1, we asked IP AI participants to write short, informal texts responding to the question: "What does the future look like for AI?" These texts were used to seed our conversations at the workshop and to give participants an opportunity to introduce themselves. Marlee Silva also conducted interviews during the workshop, which were then transcribed.

The texts provide insight into the rich set of concerns and perspectives that participants brought to the first workshop. One can see many of the concerns expressed here carried forward into the final collection of position papers, including the role and proper recuperation of traditional knowledge within technical systems; the need to protect traditional knowledge while also making (some of) it available to inform the design of these systems; the importance of language as both knowledge carrier and a primary site of computational processes; the centrality of territory in forming frameworks for understanding and communication; Indigenous communities' historical and ongoing engagement with new technologies; contesting concepts of intelligence which ignore emotional and social engagements with the world; the intrinsically cultural nature of technological systems; the cultural biases that get built into these systems;

distinctions between AI by, for and with Indigenous communities; the dangers that AI and related technologies pose towards Indigenous communities who have experienced centuries of settler colonial violence aided and abetted by the latest technologies; and the need to think about AI systems through the lenses of our specific cultures.

Articles:

I ka wā ma mua, ka wā ma hope
The future is secured by the past

Dr. Noelani Arista
February 28, 2019

My interest in AI is a continuation of the central concern of my work: that ancestral knowledge, deeply and broadly conceived will be carried over into 'the digital,' continuing into the future as it has until now; what D. Fox Harrell states is a "cultural computing perspective," which "entails performing research and practices that engage commonly excluded cultural values and activities to spur socially and critically valuable computational innovation," is an exciting concept to me. [1] In my thinking in relation to his proposition, I see how Hawaiian cultural production is held multiply as exclusive as excluded, *at the same time.*

The challenge of my work has always been how to supply access to the enormity of Hawaiian knowledges and to place them back in the everyday lives of the lāhui (the people, the nation, the community). 'The digital' poses particular challenges to the continuance of Hawaiian knowledge, in the sense that its progress doesn't leave room for the ravages which colonialism has wrought. [2]

As a historian I study the period in Hawaiian history where the technologies of the palapala (writing and print) were introduced. I have investigated how an oral/aural culture negotiated the simultaneity and transformation by, and into, the textual, how, in the 19th century kānaka maoli secured that knowledge through that transition, study which is vital to my various projects: to rebuild and understand the ontological, the epistemological, knowing and how we know, and the structures through which knowledge, story, practice, were passed on.

My research, translation and written work has focused on the training of Hawaiian intellectuals, how memories were carved (kālai ʻia) and structured to receive large amounts of data and how that data was retrieved and mobilized for particular purposes, under a regime disciplined by kapu. [3] I am studying

---

[1] D. Fox Harrell & Danielle Olson, "Cultural computing/Indigenous values," *Indigenous AI*, June 6, 2019 <indigenous-ai. net/cultural-computing-indigenous-values> 167.

[2] In my praxis language looms large as that which constructs the affective, the mode through which feeling and connection to kūpuna flows. Colonial processes hastened the loss of language and customary practice in ways that have left people with symptoms of memory loss, the inability to communicate feeling through language, and since healing was dependent to some extent upon prayer, it has given us a more difficult pathway to healing and self expression.

[3] 'Data' as in customary chant, prayer, law, history, story, some of which were quite lengthy, kept and passed on orally in a disciplined manner; and yet, these customary forms of knowledge cannot be reduced to an impersonal concept of data as unmediated by relationships. After the introduction of the printing press in 1820, many of these were re-recorded in writing and print. In addition to these new compositions moved from speech into text.

and helping to shape the transmediation of moʻolelo (history, story, authoritative speech) from textual forms into digital formats that are methodologically resonant with customary modes of transmitting knowledge. [4] I want to see these theories borne out, and I believe that Hawaiian knowledge, since we have the largest textual archive in Native North America and the Polynesian Pacific, can be an important site to contribute to what Harrell identifies in his work as an "integrative cultural system." In thinking of these systems, I am also cognizant of the limits which we in islandic communities might impose on (over)development. Several blogs have highlighted the pitfalls of colonial and capitalist tendencies trending towards extraction and consumption, and so I approach the excesses of digital formats with my desire to do what my kūpuna did, to ward knowledge (kapu), protecting it from shallow projections and proliferations which ultimately may cause lasting damage to the foundations of ʻike because of the rapidity with which incorrect, and inexact knowledge can be spread, supported, and 'shared.' Finally, I am interested in how digital formats can be Indigenized to facilitate our movement between the textual and the auditory, how to train these systems in a way that support our need to continue the passing on of our customary knowledges, histories and stories, through which the lāhui will continue to thrive.

References

Harrell, D. F. (2013). Phantasmal media: An approach to imagination, computation, and expression. Cambridge, MA: The MIT Press.

Harrell, D. F. (2019, June 6). Cultural computing/ Indigenous values. *Indigenous AI*. Retrieved from indigenous-ai.net/cultural-computing-indigenous-values.

What does the future look like for AI? : Oshkaabewis or a Skynet

Scott Benesiinaabandan

March 11, 2019

I'll answer this as it relates to my visual arts practice, involving the futurity of *Anishinabemowin* (the spoken Anishinabe language) and land/water protection and sovereignty.

Language

I think that in the near-future, AI can have an immediate impact on the preservation and promotion of endangered Indigenous languages. Already there are some projects making use of AI towards this effort. Deep learning programs designed at its root with a community's ethical concerns forming the backbone

---

[4] Moʻolelo—succession of speech acts, history, and story.

of programs can both improve research and educational resources and opportunities. Languages that are agglutinative, such as Anishinabemowin, would certainly benefit from AI driven language tools, programs that could search and scan contemporary internet resources alongside historical text archives, could provide new and intelligent responsive learning apps, driven by the particular user and their specific community contexts (dialects).

New words for new worlds is a theme I have been exploring from Anishinabemowin perspective and could see how AI assistance could provide alternate visions of the future through exploration of new language(s).

Land

Other areas where the near-future AI could be employed is in Indigenous land/water-use and sovereignty protection. Ongoing analysis of land/water-use maps could provide deeper understanding of territorial uses and importantly how best to protect on-the-land resources, such as fish stocks, forests and forest management, endangered wildlife populations, critical watersheds and high risk habitations. While drone-AI is a scary proposition, as it is mostly driven by the military and commercial interests, the same deep learning programs, coupled with the automation aerial surveillance of drone monitoring of Indigenous territories could be used as a powerful tool for Indigenous sovereignty actions.

In the Anishinabe world-view, the most important person in a ceremonial context is called *askabewis*, or "helper,." With design care and Indigenous protocols at its core, AI could be an incredibly powerful skabe working on behalf and towards the future of our communities. Seeing the opportunity in deep learning programs, and treating them as **oshkaabewis** rather than **a skynet**, is key to guiding the ethical and productive use of future AI.

ʻUmeke kāʻeo: (Re)coding AI to ʻĀina

Michelle Lee Brown

Illustrations by Kari Noe
February 26, 2019

> *ʻUmeke—bowl, poi bowl, food bowl (from the calabash gourd)*
> *ʻUmeke kāʻeo: a well-filled bowl, a well-filled mind*
> *ʻUmeke pala ʻole: calabash bowl without a dab [empty bowl, empty mind]*
> (wehewehe.org)

For the traditions I am steeped in, there is no future without the past orienting it, anchoring it, and leading it. In the title to this post, I have woven in the ideas of physical vessels - bowls, containers for the tech we use, including our own 'wetware' - with more intangible ones: minds, interconnected consciousness, vast depths of knowledge. This transference from vessel to vessel, tangible to intangible, highlights the porous boundaries between them. The title also nods to other writing I am working on around seemingly disparate sources that outline survivance as a practice of cybernetic Indigenism [1]: how Indigenous communities learn, adapt, and adjust as feedback indicates through fluid and ongoing protocols. These digital-physical, tangible-intangible materials have coded meanings within them that are routed and grounded in specific Indigenous systems, ready to (re)code and ground us.

For this introductory post, I am taking these ideas a step back—or more aptly reorienting myself to the past—by sowing seeds of something deceptively simple that will shape our futures and that of other beings and kin: the vessels we use to house these systems.

I come from salt water shorelines; we learned to navigate out, but I'm also drawn to brackish areas where fresh and salinated waters mix. Eels hatch, sharks hide, seaweed and shellfish grow rich and thick in the muck. "You find it with your nose" my Nana laughed and said. "That's where the good stuff is."

How to ensure these intelligences we help shape are well-filled ones, nourished from a mix of pasts and futures? How to take it from **AI** (as emblazoned in neon lights on Bourbon Street and some areas on the outskirts of Waikīkī) to what Noelani Arista terms [2] ʻĀIna—from illusions of fulfillment to being well-filled? These are central questions; to answer them we must use our noses to orient to the fertile (and sometimes fetid) murk of our histories. Perhaps more unpalatable: we must hear from and listen to nonhuman kin how we as human—even Indigenous ones—have taken too much (as Johnson Witehira shows in his video game *Māoriland Adventures*). Can we compost these unpleasant histories and grow? Who might help us listen and change? What kind of vessels can hold that, even when it stinks?

Stink. "No talk stink now."
When the poi bowl is uncovered, we are reminded to shift our thoughts, hold our sharp tongues.
The presence of it, in its calabash, is a reminder to come together, let go of tensions, of anger.
Stink. "It stinks, Mom."
My 3-year old daughter said this as we approached the shoreline, the wind bringing us rich Atlantic coast smells: kelp, quahog, cod and flounder that seagulls have found and picked clean. To me, it smells

---

[1] Archer Pechawis, (2014), Indigenism: Aboriginal World View as Global Protocol, in Loft, S. and Swanson, K. (Eds.) *Coded territories: Tracing Indigenous Pathways in New Media Art*, pp. 36-47. Calgary, Alberta, Canada: University of Calgary Press.

[2] "ʻĀIna is a play on the word ʻaina (Hawaiian land) and suggests we should treat these relations as we would all that nourishes and supports us." In Jason Edward Lewis, Noelani Arista, Archer Pechawis, & Suzanne Kite (2018), Making Kin with the Machines, *Journal of Design and Science* 3.5 <doi.org/10.21428/bfafd97b>.

like life. Like home. I realized how much she had to learn, how much I needed to do to (re)code her senses
when she said those words. We had been away too long.
Stink. "That STINKS!"
My comment to another adult (while our elementary-school age kids were nearby); I had just
found out I'd need to replace my phone. I'd bought my first cellphone two years before, when the
floodwaters rose in New Orleans after Katrina. I had not wanted one, but cell phone messages
were the only way my kin could communicate with us for 8 harrowing days.
Now I would need a newer one to do the work I wanted to do.
But what could I do with this one, that carried messages of hope, calls for help and of rescue,
shared our laughter and tears of relief?

I take up stink here to highlight that our technology pasts stink, and not in a good way. Past and current iterations of computers and Western communications technologies plan obsolescence into our devices, yet the housings are designed to not break down for decades, if not centuries. E-waste is being refused at recycling facilities around the world even as newer versions of devices are marketed multiple times a year. These are what roots of our ʻĀina, whether we like it or not. Smaller and larger impacts from this technology (mining, manufacturing, distribution, disposal) slide into the water we drink, the rain that falls on our crops. What futures will spring from those e-waste soils? What kind of calabash will come out of that ground? To answer this, I see two branches—two different emergences of ʻĀina.

The first will be algorhythms (rather than algorithms) that can work with older tech—cobbling it together, creating hybrid machine-kin collectives to do work for specific communities. Arthur Pechawis and Ahasiw Maskegon-Iskwew's concepts [3] of drumming across realms made me think of algorithms set to different rhythms. Technically, algorithms don't require computers (ex: geometry); an algorithm solves a problem. In Western media and computation studies, a special or highly-useful algorithm gets a name. I want to mark a category of special algorithms and name them algorhythms—these are set to different rhythms, and work with each other across digital and physical borders.

These algorhythms are Indigenously (re)coded and storied calculations and programs, ones that operate on different Indigenous community rhythms and needs; they will have their own names within communities as they build relationships with them. I also see them as interacting with other nonhuman kin, helping to address problems that occur, like the one noted by Pechawis with the Horse Nation in "Indigenism." If we take up the call to rethink what computing and technology is made of and made with, these algorhythms (and AI that emerge from them) offer rich lines of flight from what is considered castoff/outdated. This reduces e-waste and allows for groups with less re$ources to build and connect with their own systems in

[3] Âhasiw Maskêgon-Iskwêw. (1995). Talk indian to me #1. *Ghostkeeper*. Grunt Magazine Archives: <ghostkeeper. gruntarchives.org/publication-mix-magazine-talk-indian-to-me-1.html>.

Archer Pechawis, (2014). "Indigenism: Aboriginal World View as Global Protocol."

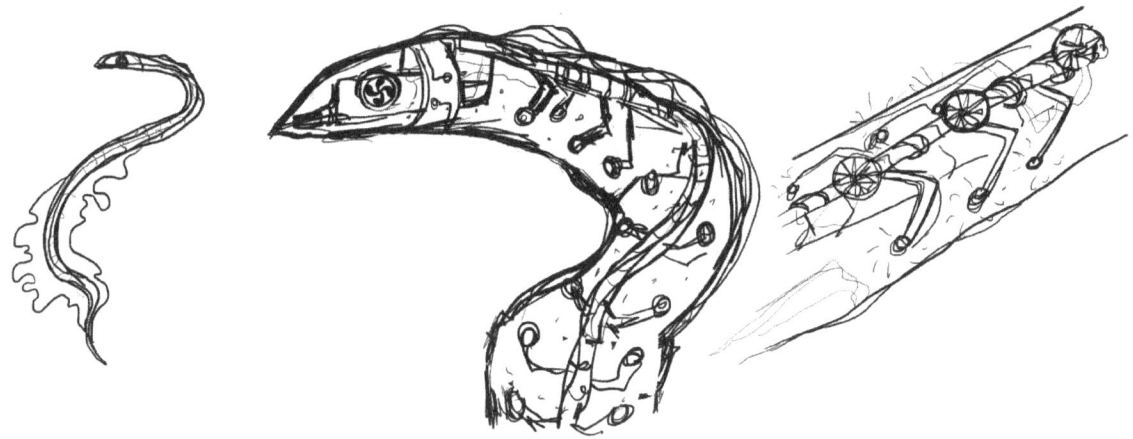

Txitxardin Lamia. Image by Kari Noe, 2019.

ways that are meaningful for them.

The second emergence is 'wetware'—biotech AI that take seriously the temporalities and materialities we are and will be. An example of this is shown below.

This is a model of a txitxardin lamia—a biotech angula/txitxardin (*elver eel* in English) that slides into specific ocean regions it is attuned to: gathering information and communicating with nonhuman relations there: algae, plankton, fish, etc. It collects and interprets this information, then enters the

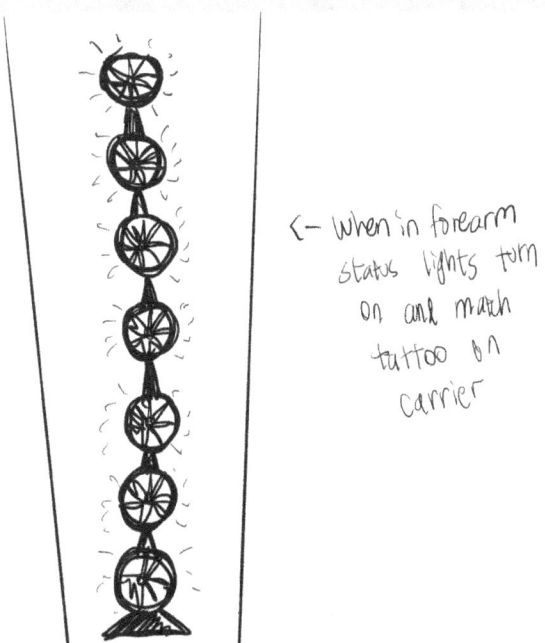

Txitxardin Lamia. Images by Kari Noe, 2019.

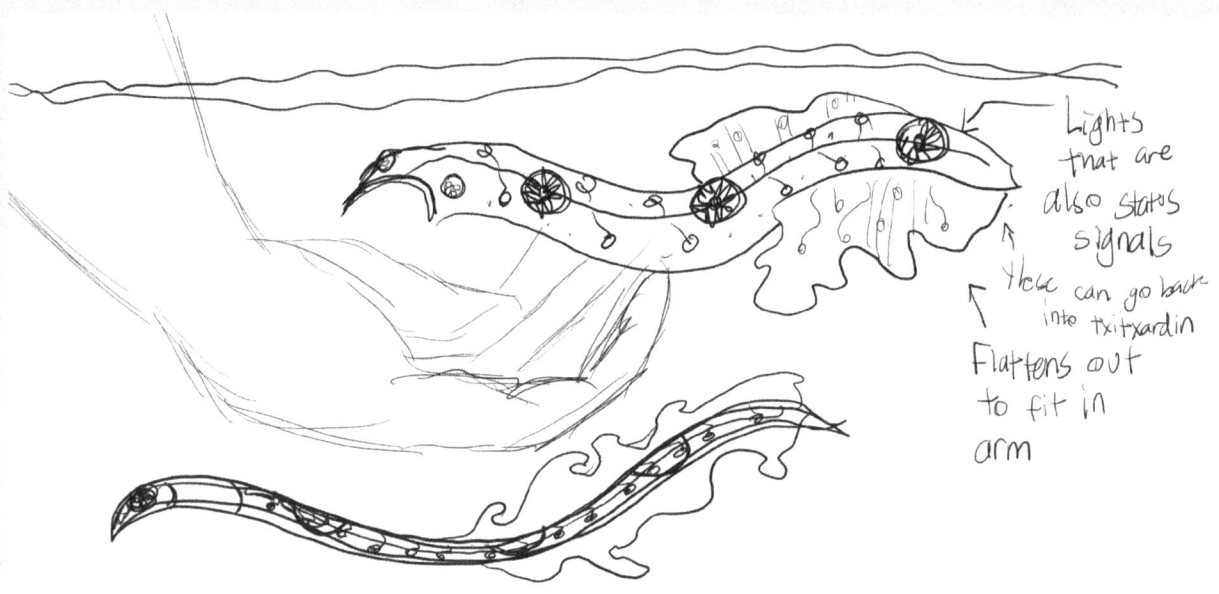

Lights that are also status signals

these can go back into txitxardin

Flattens out to fit in arm

Txitxardin Lamia. Images by Kari Noe, 2019.

arm of the sorgina (the human it works with). As it enters the sorgina and relays what it has learned, it also draws nutrients and an electrical charge from them. Each is nourished in different ways by this exchange—while intimate (and perhaps horrific to some observers), it is also consensual. This engagement is also specific to particular waters and sorginak within particular communities.

One meaning of txitxi is flesh, meat; txitxardin is our older word for eel: this is an eel made of particular kinds of flesh. Lamia (lamiak, plural) are water creatures [4] that have long assisted Basque people and received assistance from us as well - the reciprocity must be maintained. The structure of this txitxardin lamia is crucial—the casing is made from kelp and the mineral remains of the sorgina's ancestor. The DNA codes of land, sea, and AI are woven together. It is understood that the human (sorgina) will become part of txitxardin in the future, and that these beings are also temporal—they do not last indefinitely. Txitxardin are fertile and temporal vessels, well-filled as they engage with their relations; they become fertile materials for next iterations and generations to draw from as this code becomes (re)coded yet again.

There is more I could say about this example, for now I want to hold it up to highlight that the vessels we use to hold AI and algorythms matter—they shape what they do, how we connect/exchange. What can we say about the casings we use now? How would that shape if they were not 'made to last' long after we intend to relate to them and with them? If we know technological tools/kin won't always be there, how would we treat them differently?

Ideas and designs shift in provocative ways when we take up these relations as reciprocal—with elders

---

[4]   Lamiak are place-based, associated with particular rivers and streams; itsas lamiak are ocean/shoreline relations – again with very specific ares/places.

who have much to teach, nudging us to absorb as much as we can, then give of ourselves to infuse future iterations. It also means looking to our past: our e-waste past (re)coded into fertile ground for AI; our ancestral relations and recipes (re)coded as wetware and interfaces. To keep all these vessels/minds well-filled, it is important to ask <u>over and over</u> how might we give to them as they give to us. Reminding ourselves what we owe to the larger communities we are connected to—and how we negotiate those relations, what consent looks like in these intimacies.

The vessel dictates how much can be held within it—it also codes how we interact with it, how these machine-kin influence other relations on larger and smaller scales as they degrade or pile up in landfills. Wide and vast futures of potential are routed in these pasts and presents—what matters most here and now is how we (re)code ourselves internally, drawing on past technologies of relationality to structure these new kinships and the vessels that carry them.

References

Lewis, J.E., Arista, N., Pechawis, A., & Kite, S. (2018, July 16). Making Kin with the Machines. *Journal of Design and Science* 3.5. Retrieved from: doi.org/10.21428/bfafd97b.

Maskegon-Iskwew, A. (1995). Talk Indian to Me #1. *Ghostkeeper*. Grunt Magazine Archives. Retrieved from ghostkeeper.gruntarchives.org/publication-mix-magazine-talk-indian-to-me-1.html.

Pechawis, A. (2014). Indigenism: Aboriginal World View as Global Protocol. In Loft, S. and Swanson, K. (Eds.), *Coded territories: Tracing Indigenous Pathways in New Media Art* (36-47). Calgary, Alberta, Canada: University of Calgary Press.

When will computers be able to model the human brain? How will artificial intelligence impact on Indigenous communities?

Dr. Melanie Cheung, Ngāti Rangitihi

February 28, 2019

Until now, artificial intelligence isn't something I have thought very deeply about. I was never into science fiction. I didn't even take computer studies at high school. The fact that I now work in tech, despite being slightly technophobic, is kind of funny.

I am a Māori neuroscientist that has spent the best part of two decades studying the human brain from its gross anatomy right down to the molecular level. It's really an exquisite organ that allows us to see, hear, touch, taste, smell, think, feel, act, create, joke and move. Many cultures also credit the brain for having spiritual properties such as sacredness, spirituality and life force. In Māori culture, the brain is not only sacred, but human interaction with brain tissue is restricted. Consequently, I worked with *kaumatua* (elders) to develop *tikanga* (customary practices) to integrate into my scientific methods for growing cells from post-mortem human brain tissue. We continue to develop decolonizing methodologies that acknowledge sacredness, spirit, culture and community, within our laboratory and clinical practice.

My area of expertise is neuroplasticity, the brain's extraordinary ability to change its structure, function and connections in response to the input it receives. Through providing specific inputs that therapeutically alter the structure, function and connections in dysfunctional neural networks, we've been able to develop neuroplasticity-based treatments for a wide range of brain disorders. The inputs that drive these changes involve online brain training, which is how I came to be working in tech.

So, when I think about what the future looks like for artificial intelligence, I think about the increasing ability for computers to be able to model the human brain. Despite huge gains in machine learning, there are a number of limitations that computers would need to overcome to model human brains.

We know that computers can learn, so there is some degree of neuroplasticity. But computers will always be limited by the fact that they need to be programmed by humans. That is, someone still needs to program the learning. You could argue that the human brain is programmed by experience. In fact, there are several famous experiments that show that 'programming' in primate brains is actually reversible by changing input. This is the beauty of neuroplasticity.

One of the reasons that the human brain is able to change so readily is because it is a biological system. The brain contains all the cellular machinery and elements that are required for brain cell connections to be formed, reinforced and broken: DNA, RNA, proteins, neurotrophic factors, neurotransmitters, receptors, ion channels, cell membranes, energy sources, transport networks, and so on. While synthetic biologists are able to engineer artificial cells that mimic biological cells, the most complex cell they've been able to model so far is a bacterial fighting eukaryote (which is nowhere near as sophisticated as a brain cell). Artificial cells that conduct electricity, similar to brain cells, have also been created, but they are still a long way off being able to carry out the other complex cellular functions of brain cells. I think it's only a matter of time before synthetic biology and machine learning scientists combine their knowledge to develop a synthetic biology-based brain-like computer. But to what end?

Why are we interested in creating machines or robots that have human-like intelligence?

What will the real cost of artificial intelligence be on our Indigenous communities?

Will people lose their jobs because robots will be built to be more efficient than humans?

How do we develop an economy that values human qualities, as well as efficiency?

Could the vast amounts of money poured into artificial intelligence research be better used on improving Indigenous health and living conditions or protecting our environment?

How then, can we develop artificial intelligence technology that improves quality of life for Indigenous people rather than creating yet more disparity?

How can Indigenous people be involved in decision making about artificial intelligence?

What do our elders have to say about artificial intelligence?

Will Indigenous voices be valued in this space?

What decolonizing methodologies can we develop to determine how we want to interact with artificial intelligence?

In summary, I am looking forward to our workshop together. I'm excited to meet Indigenous people from diverse disciplines. I'm especially looking forward to spending more time thinking deeply about artificial intelligence and the ways it might impact on Indigenous communities in both good and bad ways.

What does the future look like for AI?

Meredith Coleman

18th Feb 2019

Artificial Intelligence already surrounds so much of what we do in our day to day lives—for example self-service scanners in supermarkets were posited as a 'creepy futuristic machines' when they were first introduced in the mid-noughties, yet these are now a much-appreciated convenience for shoppers, and asking Siri or Alexa rather than typing a question into Google has become second nature to many. Shaving a few seconds from one's day has become preferable in many cases to maintaining our privacy, willingly giving our precise location and other personal details to companies such as Google, Facebook and Uber in the name of convenience.

We are already living in the future, in some respect, as much of our technological progress becomes focused on refining what we have already created—although perhaps this is a naïve view from someone who can't picture how different the future may really look. Today's world looks vastly different from the world of the 1990s, for example, except that we still use much of the same technology. Might it be the case that twenty years from now, artificial intelligence and technology are aesthetically very different, yet their function remains similar? Might we be using the same basic technology for brain surgery that we've used for years, while the success rates and accuracy of the technology dramatically improve?

In Kate Darling's TED talk on our emotional connection to robots,[1] she raised questions about why, as

[1] Kate Darling, "Why we have an emotional connection to robots," *TED*, September 2018, <ted.com/talks/kate_darling_why_we_have_an_emotional_connection_to_robots>.

humans, we seem to feel emotion for certain technology as though it were alive. I think this is important when considering where the future of AI will take us, in particular as Darling raises issues of what happens when humans are unable to disconnect from technology emotionally. It may be the case that the more specialised and progressive our technologies become, the less we are able to separate ourselves from them emotionally. Darling spoke specifically about robots being used to clear minefields, and other army robots even being given funerals when they were "killed,." In light of this week's news that the Mars rover "Oppy" Opportunity has 'died', this emotional connection seems to have really hit home, as we have seen the direct impact that an emotional connection with robots and technology can have.

But perhaps this is a good thing. Does this not show us that humans are not so desensitised to violence and destruction, to the degree that we will mourn for something that is not even alive? Darling's talk highlights for me how humans are still very much in touch with our emotions, and we seem to be a long way off being made robotic ourselves in our inability to care. One of the greatest worries for the upcoming generations is that an increasing demand for artificial intelligence will result in humans being less reliant on other human company, as the need to communicate with one another is stripped away by technology. Darling's research suggests that this is not the case, at least not yet, as our ability to empathise still outweighs the abilities of the technology we have created. While it remains true that the technology that exists today is capable of doing terrible things, it simultaneously seems that to most people, improving on technology is largely for positive progress. Yes, artificial intelligence is reducing our need for learning certain skills (think being able to have food delivered through our phones and the internet, rather than learning to cook for ourselves), yet these same technologies can help us to learn skills we might not otherwise have the opportunity to explore—for example devices such as Alexa and the Google home hub being able to use the internet to create walkthrough instructions for people to learn as they go. I mentioned earlier that people are becoming increasingly fond of convenience, and it seems to be the case that the progression of technology and artificial intelligence is most appreciated when it allows the user to add a level of convenience to their lives, rather than having our lives be taken over by the reach of artificial intelligence. In particular, technology has practical uses in the disabled community, from screen readers for accessing social media, to the specialised treatment of disease. Being able to harness new technologies to aid specific groups opens doors for creating a more accessible society for all.

Overall, it seems that the future of AI is incredibly bright, with new technologies being produced on a near-constant basis. While popular culture increasingly prophesises how artificial intelligence will be used for the downfall of civilisation (dramatic, but perhaps not too hyperbolic), with the likes of Elon Musk becoming the comedy villains of our real-life superhero movie, it seems that we are far from being taken over by a robot race. It is inevitable that artificial intelligence will become a much larger part of everyday life in the coming years, however this does not need to be "the escalator from hell," [2] as Jack Clark, the head of policy at OpenAI worries that their latest AI technology will become if released to the

---

[2]   Alex Hern, "New AI fake text generator may be too dangerous to release, say creators," *The Guardian*, February 14, 2019 <theguardian.com/technology/2019/feb/14/elon-musk-backed-ai-writes-convincing-news-fiction>.

public. These concerns surrounding AI are not entirely without reason, with privacy and data breaches being front and centre of many news stories in recent months, however it seems to be the case for now that much of the technology for now is being used for public good—even if vast quantities of personal data are being stored by corporations. It is difficult to say whether AI will ultimately have a wholly positive or negative impact on society, since so much of the technology is being created and worked with while not necessarily being fully understood. We are at a point in history where science is progressing at an incredibly fast pace, with new concepts being realised constantly, as predicted in the 1960s by Gordon Moore. Working with such technology means that fundamentally, we are not fully equipped to deal with the full extent of its capabilities. The coming years are likely to bring a massive change in how we interact with the world around us, as well as with one another, and may exact immense social change around the globe on a much larger scale. It is impossible to say whether Jack Clark's concerns or Kate Darling's optimism for the future of AI and technology will become the realised state, but with the rate of progression it seems sensible to accept that either approach is a distinct possibility for our future.

References

Darling, K. (2018, September). Why We Have an Emotional Connection to Robots [Video file]. *TED*. Retrieved from ted.com/talks/kate_darling_why_we_have_an_emotional_connection_to_robots.

Hern, A. (2019, February 14). New AI Fake Text Generator may be too Dangerous to Release, say Creators. *The Guardian*. Retrieved from theguardian.com/technology/2019/feb/14/elon-musk-backed-ai-writes-convincing-news-fiction.

Envisioning the Artificial: Technology, Time, and Indigenizing The Future of AI

Ashley Cordes

November 19, 2019

As tired as it is to say, thanks to the *Black Mirror* Netflix series, the music of Janelle Monáe, psychedelia, sci-fi, and a plastic bag full of cultural artifacts, the pop cultural psyche already has a clear collective visioning of what the future will look like via Artificial Intelligence (AI). How do we move beyond what is already semiotically pre-determined to ask the negotiated and oppositional ways that the future looks like *for* AI?[1] We need to question what the future looks like for AI, because we and AI are among the

---

[1] See Hall, S. (1973). Encoding/Decoding. *In Culture, Media, and Language: Working Papers in Cultural Studies, 1972-1979* (pp. 128-138). London, UK: Hutchinson.

many agents determining it. In taking these steps, there are challenges in theorizing what technology of the future is because of the work that theorizing it performs within an overarching capitalist, sexist, and racist system.

In the '90s particularly, the popularization of digital was framed by discourses of transformation, replacement, and advancement (Van den Boomen, 2009). Digital culture came to define the zeitgeist, slighting the so-called generalized print and electronic eras that preceded it. Maintaining the pretense of these eras as separate and linear is but one tactic used by tech industries to sell their newer products because technologies are on one hand commodities not gods—and on the other hand, gods.

These narrative of progress tends to help humans more generally by making them feel comfortable about their movement through time and space and they hold profound social meaning. Most relevant is that the narrative of progress reflects racially charged ideologies that become hyper-naturalized. As Mètis critic Emma LaRoque (2010) states, "behind the dichotomy of civilization versus savagery is the long-held belief that humankind evolved from the primitive to the most advanced, from the savage to the civilized" (p. 39). They blur the fact that communities, particularly Indigenous communities, have been using many technologies, shifting, retaining, rearticulating, and adopting different forms for tens of thousands of years or since time immemorial.

In this regard, history and time is too commonly described as existing on a horizontal line with the far left being the past and the right being the future within certain worldviews. The narrative of progress creates a laughable spectrum that tends to place Indigenous technologies (old and/or new) in the past and uber-new new media on the right, despite Indigenous spirals upon that slippery spectrum and their/our clear contributions to the uber-new new media. I, before recently reading disappointing writing on Indigenous currency technologies (see the forward of _Paid: Tales of Dongles, Checks, and Other Money Stuff_, which describes 'shell money' as weird, depressing, and non-modern (Sterling, 2017, x-xi), thought that this was already clear. The only thing that is now clear is that conversations like these need more space.

Specifically some space needs to be focused on the future of AI because it is now being paired with almost all other preeminent technologies. AI will infiltrate privacy while simultaneously being personal assistants, fuel technology races between nations and be adopted for warfare while still driving you home from work. It will, as it already has, come under large scale scrutiny and regulation for the bias it inherently holds when used in criminal justice, healthcare, lending, and education. It will also look hopeful. The future will be techno-pessimistic, optimistic, and pragmatic and it's not productive or holistic to look at in only one way. Moreover, looking at it in only one way is a means of forcing arguments that we just simply are not sure of and limiting a categorically more creative visioning of the future of AI.

What makes newer technology, AI, or media interesting, meaningful, and worthy of talking about is when technological innovations are thought about by communities that have been consciously marginalized by the system. The innovations themselves are not necessarily paradigm shifts, but the ways in which the systems created are commandeered to change up the systems in some small way are. Foucault, a critical theorist, discusses how power is, of course, an omnipresent feature of life. One side or many sides of a power equation pushes the others in a direction and visa-versa, a tug-of-war of sorts. Power is not only everywhere and two-directional but is targeted, enacted, and embodied through discourses and knowledge. The discourse that centers on Indigenous people as technologically backwards is one deployed by colonial forces to delegitimize Indigenous ways of knowing and ways of acting, and it is disconcerting that we still have to talk about it. By making AI work with us and framing it as an Indigenous project, efforts like these play into a chipping away at this 'regime of (un)truth.'[2] They break down epistemological underpinnings and exemplify the fact that Indigenous people are not only surviving in the digital age, but are in the driver's seat of envisioning futurity in an increasingly digital and globalized world.

Technology, the communicative artifacts that are considered in the deployment of stereotypes such as these, are at the same time the essences that can be re-inscribed or created with counter-hegemonic charge. With Indigenous efforts the future of AI will feel like predicting, planning, learning, representing, executing, doing, perceiving, solving, fixing, ruining, helping, hurting, intellectualizing, complicating. The ride will not look like a linear line and it will also recognize and give nods to glimpses of AI in 'traditional' items. For example, Haas (2007) points out that hypertext and multimedia are too often claimed as Western. Hypertextuality refers to the accessibility of texts through other texts, layered with meaning. Wampum shells were, and still are, made by many Indigenous peoples, particularly Haudenosaunee peoples, into intricate 'belts' to tell stories, to mark occasions, to make contracts; there are layers of meaning that make them hypertextual. These are also arguably digital in that the beads are strung, they are code, and can be read; they are retrievable, decodable, memories of Indigenous epistemology. This suggests an intelligence in the creation of life that carries on beyond when human and non-human 'creator's' hands have left said technology.

AI can look to help make better the lives of Indigenous people and help to ensure Indigenous futurity. For this to happen, AI should be made/stewarded with Indigenous epistemologies at the forefront to radically question, appropriate, and push back pervasive globalized peer-to-peer systems or any systems which may not help our communities. Here lies the potential to help restructure our social worlds, transform the ways we view digital territoriality, and help us to embody relationships with the various ecosystems we depend on.[3]

References

Haas, A. M. (2007). Wampum as hypertext: An American Indian intellectual tradition of multimedia theory and practice. *Studies in American Indian Literatures*, 19(4), 77-100.

[2] See Foucault, M. (1978). *Discipline and punishment: The birth of the prison.* New York, NY: Pantheon Books.

[3] See Wildcat (2009) for more on ecosystems, environmentalism, and Indigenous knowledge. Wildcat, D. R. (2009). *Red alert!: Saving the planet with Indigenous knowledge.* Golden, CO: Fulcrum Publishing.

Foucault, M. (2002). *The Archaeology of knowledge*. London, UK: Routledge.

LaRocque, E. (2011). *When the other is me: Native resistance discourse, 1850-1990*. Winnipeg, MB: University of Manitoba Press.

Sterling, B. (2017). Forward. In B. Maurer & L. Swartz (Eds.), *Paid: Tales of dongles, checks, and other money stuff* (ix-xii). Cambridge, MA: The MIT Press.

Van den Boomen, M. (2009). *Digital material: Tracing new media in everyday life and technology*. Amsterdam, NL: Amsterdam University Press.

A very personal look at the future of AI

Joel Davison

AI today is bound by practicality, talented developers, cutting edge research, specialised hardware and top of the line cyber security, which are all ingredients required to advance simple AI beyond current offerings. This means that the entities with the power to advance AI, those with access to pools of talent and academic connections as well as the funding for hardware and security, are those which already have much more money to invest than what is required to operate as a business. These entities, be they government or private, expect a return on investment, in this way AI advances will always be pushed in a direction that is either profitable or marketable, due to this AI is entwined with automation in our cultural lexicons and it is this connection that often dominates conversation.

If Artificial Intelligence is to replicate human intelligence, then the most direct way to profit off of said intelligence is to exploit its labor value. In this way conversations are often steered towards analysis of labor-value of existing occupations. For example, advances by large tech companies in self-driving cars has every in-tune truck driver eyeing other industries at this point, and we [1] can't [2] stop talking [3] about [4] it. [5]

[1] Walker Orenstein, "Automated 'platoons' of trucks might soon be driving on Minnesota roads," *MinnPost*, February 1, 2019 <minnpost.com/good-jobs/2019/02/automated-platoons-of-trucks-might-soon-be-driving-on-minnesota-roads/>.

[2] Seth Clevenger, "Self-driving truck startups TuSimple, Ike attract more investment to fuel development," *Transport Topics*, February 13, 2019 <ttnews.com/articles/self-driving-truck-startups-tusimple-ike-attract-more-investment-fuel-development>.

[3] Adam Rowe, "The trucking industry's future: go high tech or go home," *Tech.Co*, August 30, 2018 <tech.co/news/trucking-industry-future-autonomous-drivers-vr-2018-08>.

[4] David Welch, Gabrielle Coppola, & Chester Dawson, "Young CEO of electric vehicle startup Rivian has Amazon riding shotgun," *Seattle Times*, February 24, 2019 <seattletimes.com/business/young-ceo-of-electric-vehicle-startup-rivian-has-amazon-riding-shotgun>.

[5] Finn Murphy, "Truck drivers like me will soon be replaced by automation. You're next," *The Guardian*, November 17, 2017 <theguardian.com/commentisfree/2017/nov/17/truck-drivers-automation-tesla-elon-musk>.

The vast majority of these industry shaping moves that are being made are opportunities presented only to the wealthiest organisations on the planet, due to the benefit only being realised at a huge scale thanks to the costs outlined above, talent, research, hardware and security. It simply isn't feasible for small organisations, potentially social ventures, NGOs or co-ops, to lay stake to a portion of the market without the network and capability to take advantage of the wider market. If the benefit of Artificial Intelligence in this liberal-capitalist frame is the profit earned by extracting more labor-value by reducing the overhead of hiring humans to manually perform tasks, then by the time you have paid the up-front costs for the research, development and specialised manufacturing to begin providing self-driving vehicles as a service, you start to realise that you need to roll out your service on a massive scale to begin to realise the benefits. In this environment Artificial Intelligence becomes a winner takes all venture, where the only participants are those already winning.

However, we have been seeing a shift in this landscape, a move by some of the largest organisations that changes the climate entirely. Having developed their AI and taking their time to scale and implement before they start to see their benefit, these large organisations have started to look for alternate revenue sources for their AI solutions. Most notable of these alternate revenue sources are the AI as a service platforms, such as IBM's Watson or Google's Tensor Flow. Suddenly, small organisations can provide the benefit of AI (or at least market that they do) without the tremendous up-front cost of research and specialised hardware. In this we are now seeing many small businesses and startups getting into the game of exploiting the difference in labor value between human intelligence and Artificial Intelligence, this time opening up smaller scales, nooks and crannies in the marketplace to be explored.

In all of these conversations we are only exploring the capital value of simple Artificial Intelligence: it's the capitalist equivalent of only talking about the 'why?' of AI (the answer to which is almost always 'money'). Little do we explore the impact of simple Artificial Intelligence, we never really ask 'how?', and when we do it's always too late.

In November 2017, The Guardian broke the story of a secret police blacklist employed by the New South Wales Police,[6] a "Suspect Targeting Management Plan," which the NSW Police Commissioner called a "predictive style of policing,." This is kind of low-hanging fruit isn't it? My intention was to share a couple of cases where organisations hadn't stopped to ask 'how?', or what their impact is, but surely no one on this program even stopped to ask 'why?'. It doesn't take a genius to figure out how this goes terribly wrong, hell you don't even have to look much further than Marvel, who ran a (fantastic, by the way) crossover event by the title of "Civil War 2" which featured at its center the arguments for and against 'predictive policing', it's actually kind of prophetic and I love it so.

---

[6]   Michael McGowan, "More than 50% of those on secretive NSW police blacklist are Aboriginal," *The Guardian*, November 10, 2017 <theguardian.com/australia-news/2017/nov/11/more-than-50-of-those-on-secretive-nsw-police-blacklist-are-aboriginal>.

spoiler warning

The event comes to boiling point when a new Spiderman, <u>Miles Morales</u>[7] (A young African American, Puerto Rican man) is accused of murdering Steve Rogers, Captain America in the future. After all of the superheroes have shared their perspectives and opinions and had their brawls, the takeaway from this is the question, 'is it ever okay to judge someone for something they haven't done but could do?', to which the answer is no, you shouldn't, especially if the current criminal justice system is suited to it and especially if you don't think very carefully about it. Unfortunately the Australian criminal justice system isn't suited to it and very clearly the NSW police did not think very carefully about it.

spoilers over

'Okay Joel so you have some comic-books-based opinions on predictive justice, but seriously how bad could it be?'

It gets pretty bad. According to the NSW Police Commissioner Mick Fuller, "here were about 1,800 people subject to an STMP across the state. About 55% of them were Aboriginal," the youngest of which is a nine year old. Currently Indigenous Australians only make up 3% of the national population, so how is it that we represent such a large portion of this database? Are we really that talented at crime? I mean, do we really commit 17 times more crime than any other Australian ethnicity? Of course not, that's ridiculous, so how did this AI come up with this list of suspects? The truth is, we don't know and if you ask the police they wouldn't know either, the company that they contracted to develop the solution likely don't know either and don't care how, they've already answered their 'why?' (read: money). Most likely the people developing the solution don't understand how the AI's learning algorithms work and didn't think about the kind of training data the AI was trained on before it started working on production data.

'But Joel, they'd have to have thought pretty hard if they made the AI racist, it's a machine so it's impartial to race and ethnicity', <u>turns out that's not the case,</u>[8] AI more or less come out of the box as racist. This is due to how AI are configured in these projects, to perform better than humans they need to learn more than humans in the narrow field they're being developed for, which is one of their strengths: they can take a huge set of training data and learn from it very quickly. The data is important, however, and as it so happens the most easily accessible large datasets are user-generated and contain all of their respective prejudices. So it's important to ask 'what data set was it trained on?', in this case definitely existing data on previous arrests and criminal convictions by the Australian Federal Police. 'Hold on, the data on previous arrest and criminal convictions by the Australian Federal Police reveals a strong recurring prejudice toward the Indigenous population of Australia?'

---

[7] "Miles Morales (Earth-1610)," <marvel.fandom.com/wiki/Miles_Morales_(Earth-1610)>.

[8] Robyn Speer, "How to make a racist AI without really trying," July 13, 2017 <blog.conceptnet.io/posts/2017/how-to-make-a-racist-ai-without-really-trying>.

Imagine my shock.

So now the police have a racist AI that's populating a confidential list of suspects who are majority Indigenous, who the police are now legally able to arrest before they commit a crime or do anything suspicious. Yeah, the police in 2017 criminalised being Aboriginal. That's how bad it gets.

I'd love to say this proves the point I was making earlier about the impacts AI can have if we don't ask 'how?' but it's even worse than that. The fact of the matter is unless we are very careful, AI-as-a-service can be used to intentionally obfuscate the 'how?'. We don't know how the NSW police's AI became a racist, we can make very good educated guesses about training data and configuration, but we don't *know*: the AI obfuscates the process by which it came up with its database through its sheer complexity alone. The biggest problem is that in spite of this, the results are still being used with authority. Because it is an AI, a machine that 'just runs analysis' all it is doing is giving authority to existing and past prejudices and perpetuating said prejudices, rather than having the ability to challenge them like a human might.

We haven't been asking of ourselves 'how?' and when we don't, we don't move forward, we don't challenge and we don't change. We just become more efficient and I don't think that's the vision anyone who is passionate about AI & Computer Science imagine. If we are to use AI to move our society forward, to make real change instead of just making profit, we need to ask 'how?'.

References:

Clevenger, S. (2019, February 13). Self-driving truck startups TuSimple, Ike attract more investment to fuel development. *Transport Topics*. Retrieved from ttnews.com/articles/self-driving-truck-startups-tusimple-ike-attract-more-investment-fuel-development.

McGowan, M. (2017, November 10). *The Guardian*. Retrieved from theguardian.com/australia-news/2017/nov/11/more-than-50-of-those-on-secretive-nsw-police-blacklist-are-aboriginal.

Miles Morales (Earth-1610) [online wiki page]. (n.d.). *Marvel Database Fandom Wiki*. Retrieved from marvel.fandom.com/wiki/Miles_Morales_(Earth-1610).

Murphy, F. (2017, November 17). Truck drivers like me will soon be replaced by automation. You're next. *The Guardian*. Retrieved from theguardian.com/commentisfree/2017/nov/17/truck-drivers-automation-tesla-elon-musk.

[Online article]. (2019, February 24). Retrieved from pressreviewer.com/2019/02/24/the-leading-companies-competing-in-the-global-mining-truck-market-industry-forecast-2018-2022.

Orenstein, W. (2019, February 1). Automated 'platoons' of trucks might soon be driving on Minnesota roads. *MinnPost*. Retrieved from minnpost.com/good-jobs/2019/02/automated-platoons-of-trucks-might-soon-be-driving-on-minnesota-roads.

Speer, R. (2017, July 13). How to make a racist AI without really trying [Blog post]. Retreived from blog. conceptnet.io/posts/2017/how-to-make-a-racist-ai-without-really-trying.

Rowe, A. (2018, August 30). The trucking industry's future: go high tech or go home. *Tech.Co*. Retrieved from tech.co/news/trucking-industry-future-autonomous-drivers-vr-2018-08.

Welch, D., Coppola, G., & Dawson, C. (2019, February 24). Young CEO of electric vehicle startup Rivian has Amazon riding shotgun. *Seattle Times*. Retrieved from seattletimes.com/business/young-ceo-of-electric-vehicle-startup-rivian-has-amazon-riding-shotgun.

Cultural Computing/Indigenous Values

D. Fox Harrell, Ph.D. & Danielle Olson

February 2019

Artificial intelligence (AI) systems are cultural systems. This may not seem intuitive for those who think of them as complex technologies serving utilitarian purposes. However, "all technical systems are cultural systems" (Harrell, *Phantasmal Media*, p. 345). This is because technologies are created in particular historical-cultural contexts and are informed by underlying shared cultural perspectives. Furthermore, computers play a role in shaping culture "through facilitating the construction of shared knowledge, shared beliefs, and shared representations" (Harrell, 2013, p. 345). When considering the future of AI, particularly the relationship between a plurality of Indigenous values and AI, we need to then make some of the values within AI explicit that are usually left implicit. Toward this end, it is first useful to consider what AI itself is—and we quickly begin to see that AI itself represents a plurality of values as well.

AI represents many different aims, technologies, approaches, and communities of practice. Often times, these aspects are described in binary terms, for instance contrasting:

CONTRASTING FEATURE	SIDE A	SIDE B
Aspirations	Strong AI[1]: Aspires to machine consciousness, sentience, etc.	Aspires to competence in a more narrow domain (e.g., performing indistinguishably from humans in conversation)

[1] Searle, 1980

CONTRASTING FEATURE	SIDE A	SIDE B
Approaches	Symbolic: a.k.a. "Good Old-Fashioned AI," (GOFAI)[2] Uses high-level, human- readable representations (e.g., first order logic)	Connectionist: Uses artificial neural networks as a model
Research Goals[3]	Engineering: Produce a system that performs some task typically thought of as requiring intelligence	Cognitive Science: Produce a system that helps explain or simulate human mental or neural processes
Style[4]	Neat: Preferring top-down explainable, if not provable, solutions	Scruffy: Preferring bottom-up, functional, if not completely explainable, solutions

Support for AI has gone through cycles as well. Early on, AI researchers worked on abstract, small domains with the belief that the results would generalize to the world at large—with a swath of research impelled by military-industrial applications. The mid-1970s have been described as an "AI winter," particularly in the United States as the Defense Advanced Research Projects Agency (DARPA) funding policy changed in a way that disadvantaged generalized AI research. Recently, with the processing power of today's computers, pervasiveness of big data, and new innovations and optimizations with artificial neural networks, 'deep learning' approaches have produced compelling results. The attendant attention and funding AI have prompted some to even suggest we are now in an "AI spring" (Warren, 2016).

In light of these many aspirations, aims, approaches, research goals, and styles of AI, one might ask: how might we begin to characterize the values within traditions of AI? The concept of an 'integrative cultural system' helps toward this end. The term 'integrative cultural system' can be used to describe how culture, knowledge, beliefs, and representations are distributed onto material and conceptual artifacts, here with a focus on computational artifacts (Harrell, 2013, p. 207-249). We need to carefully examine the assumptions, structures, uses, discourse around, and practices involving these technologies. This means that we should not limit ourselves to analyzing the technical functionality of systems, but rather looking at the ecologies of people, artifacts, code, interfaces, language, etc. around systems in a more holistic way (Harrell, 2013, p. 74).

[2] Haugeland, 1985

[3] Jenson, 2018

[4] Schank, 1983

Finally, to engage the relationship between Indigenous cultures and AI in a manner that supports people's empowering needs and values, we need to adopt a cultural computing perspective (Harrell, 2013, p. 167). This perspective means entails performing research and practices that engage commonly excluded cultural values and activities to spur socially and critically valuable computational innovation. More importantly, cultural computing research and practice focuses on rigorously understanding and articulating the groundings of computing systems in culture. This all means that we must work together to build the future of AI in a manner that supports the vast array of human creative cultural production, including supporting mental and physical wellness, economic and educational advancement (U.S. Global Development Lab, 2018), the arts, and more. We hope that this workshop can help open new vistas based on grounding computational practices in Indigenous values that have long traditions of supporting such ends.

References

Harrell, D. F. (2013). *Phantasmal media: An approach to imagination, computation, and expression.* Cambridge, MA: The MIT Press.

Haugeland, J. (1985). *Artificial intelligence: The very idea.* Cambridge, MA: The MIT Press.
Jensen, G. (2018, June 10). Artificial intelligence: cognitive vs. engineering approaches [Blog post]. Retrieved from gavinjensen.com/blog/cognitive-vs-engineering-ai.

Schank, R.C. (1983). The current state of AI: one man's opinion. *AI Magazine* 4(1), 3-8. doi.org/10.1609/aimag.v4i1.382.

Searle, J. (1980). Minds, brains and programs. *Behavioral and Brain Sciences* 3(3), 417–424. doi.org/10.1017/S0140525X00005756.

US Global Development Lab, Paul, A., Jolley, C., & Anthony, A. (2018, September 5). Reflecting the past, shaping the future: Making AI work for international development. *USAID.* Retrieved from usaid.gov/digital-development/machine-learning/AI-ML-in-development.

Warren, C. (2016, May 20). Google's artificial intelligence chief says 'we're in an AI spring'. *Mashable.* Retrieved from mashable.com/2016/05/20/google-ai-spring/#zpmxv2hfNGq3.

Digital Sovereignty

Peter Lucas Jones

June 18th, 2019

Kia ora, my name is Peter-Lucas Jones and I'm from Te Hiku o te ika, and my iwi, or my tribes, are Te

Aupouri, Ngai Takoto and Ngati Kahu.

So I met Oiwi Parker Jones at Oxford University, which was in year 2018. We had the opportunity to meet with him and a few of his colleagues and talk about Maori language voice recognition and the opportunities that that gave to our people to actually synthesize a voice in our language, to actually develop a huge text corpus for, a data for development and innovation and along with that an acoustic database or an acoustic data collection with all the reading of utterances in our language in order to develop natural language processing tools.

So that's how I met Oiwi and then he contacted Keoni Mahelona, who is my partner, and then that's how I got here. Yeah, after writing a couple of paragraphs around what I could possibly bring to the table, knowing that this is about participation but also sharing our expertise, experience and what knowledge and skills that we bring to complement designing a solution for a problem we all share as Indigenous people.

I'm often mindful that we get invited as Indigenous people to Indigenous conferences that are organized by non-Indigenous people. So my expectation was that this was being organized by Indigenous people, so I think the level of participation that you're quite happy to be part of when it's a hui, or a workshop or a meeting that is organized by people that are from communities similar to yourselves, then you think, well they understand the context of colonization, white assimilation in a in socio-economic background and place that affords us as Indigenous people, and quite often it's at the bottom of the heap. So I was looking at it as a way to secure a place for ourselves, my tribe, my tribes, you know—my people, in the future, in the digital future.

Because as far as I'm concerned I don't just think about how AI can be used, I think about how we can be the makers of AI and how do we secure economic opportunity for our people in the future, so that when we deal with open source and all that that offers us, we deal with it with our eyes wide open, knowing that these are the skills and expertise that we need to apply open source code or whatever.

Let's face it: most of our people are not in a position of privilege that affords them those skills and expertise, so open source is good for white privilege.

But what does open source amongst Indigenous communities look like? How do we share ideas, concepts with a level of integrity and trust that you only have with other Indigenous people?

When we look at the artificial simulation of human intelligence we're mindful that that operates a great deal of the time on the data that it is fed for training, computer modeling and all that type of behavior that we expect it to perform relatively well at.

If we look at the jail. For Maori people, we make up more than 50% of the jail population yet we are only 15% of the wider population.

Quite often a reason for that is described as racism or racial profiling but if we look at the other Pacific peoples that are in our wider population in Aotearoa, New Zealand, we can see that only 11% of the jail population is actually made up of other non-Maori Polynesian or Pacific Island people. So then that suggests something quite different.

We know that most of our people at least have a first or second degree relative that has either been to jail or is in jail. So when we talk about law enforcement and AI we're mindful that, what are the risks there that we need to be mindful of. Data, if it's being mined or if it's being categorized or if it's being curated in a way for law enforcement, needs to take into account that that data is biased.

We know that white people get let off for crimes that our people get sent to jail for and so that's a risk that we've identified. But along with it comes, along with AI, comes a lot of opportunity.

If we were to think about natural language processing tools, if we were to think about the important part that we place on language retention and the acquisition of our languages and our culture, we know that that sort of data is captured in our written text. It's also captured in the stories that we tell, intergenerationally amongst our people through speaking our language.

So if we were to synthesize a voice or if we were to develop voice to text, text to voice, or even voice to voice, what does that open up in terms of opportunity for cultural and language intergenerational transmission in today's day and age?

So whilst there are risks, we can't run away from the opportunities. Because as people that have been alienated from our culture, we now have an opportunity to sometimes revive things that we have lost as part of the colonization process.

So I think working with other Indigenous people that have similar problems, we come up with a solution or a series of solutions that we can then pick from, knowing that we trust other Indigenous people because they've gone through a similar traumatic experience to ourselves.

I mean, imagine if we could automatically transcribe Maori language audio in real time and the traditional knowledge we could unlock from there?

Our extensive native speaker collection that we have at our iwi radio station, I'm the general manager for my tribal broadcasting media hub, if you think about all the traditional knowledge, the medicines, the foods. We talk about food security, we talk about restoring the water ways, what sort of plants grew down a specific water way. We talk about the ocean, we talk about the mountain, we talk about the forest, the birds and all the animals that are part of our landscape. And when we think about natural language processing tools and using that as a way to mine our own data for the purposes of revival, maintenance, preservation, promotion and growth of our language and culture, it opens up so many amazing opportunities and that excites me.

I think that we naturally gravitate towards people that have shared problems and what I'd like to see come out of this is us to be able to at least group our shared priorities.

I'm very optimistic in terms of what we can achieve and you can hear that we're talking about the environment, we're talking about our landscape, we're talking about our language, we're talking about our culture. We're talking about data security and data storage.

We store our data in our song and dance. We store our data in the way that we cook. We store our data in the way that we perform oratory. We store our data in the way that we welcome people and we store our data in the way that we farewell people.

But in the modern age how are we going to store our data being mindful that we do not live like we traditionally used to?

I grew up with our grandmother, our grandmother's sisters, our uncles and aunties, our mother and father. Our cousins were like our brothers and sisters. But now our families are growing up with a mum and a dad in a western context. So how can we use artificial intelligence to simulate the way in which our families are connected and the way that we transmit inter- and intra-generationally? Because I think that's a big part of our shared problem, is we are now displaced from the places we are most connected to.

So how do we reconnect ourselves without observing the community and starting to participate in it?

I think that we've got to be mindful that we have to enable development and innovation. We should be protective of our data, we have every right to be. We have a responsibility to protect our data. But with the protection also comes the role to promote and grow and we cannot promote and grow if we are going to constantly live in fear.

So I think what we have here today, and yesterday, is a group of people that are ready to risk it all, and we know that people that are ready to risk it all are going to be leaders.

They're going to be leaders that take these concepts and new ideas back to our communities so that we can take hold of these opportunities and when we do that we know that we're going to be moving with our brothers and sisters. And I think there is a level of security and when we can offer that back as a report to the communities, the Indigenous communities that we come from, we can then seek the ongoing endorsement and support. Because it's not about us making the decision on behalf of our people, it's about us taking these ideas back to our people and seeing if they're ready to engage.

And I think that the time is now and I think that this workshop couldn't have brought together more passionate people that are related and very entrenched in their own Indigenous communities and development and innovation, cultural and language preservation and very much connected to the landscape.

So I'd just like to say kia ora, thank you for inviting me, but most of all thank you for allowing us to share and receive, of course, the offerings from our brothers and sisters from other Indigenous parts of the world. Kia ora mai ano tātou.

What does the future of AI look like?

Kekuhi Kealiikanakaoleohaililani

February 28, 2019

> *from the lowland forest of Panaʻewa*
> *our aloha to you of the makaliʻi constellation*
> *to you of Kānehoalani-sun and his fiery volcanic offspring*
> *our greetings to you of the oceanic people*
> *to you of the mountainous regions*

OUR fondest aloha to you all as you traverse Kanaloa's vast ocean memory on your winged canoes and finally land on our ocean-bound island home...Hawaiʻi! Welcome home.

Thank you for the posts ahead of mine. In addition to very little inquiry on my part, your contributions are helping me form some understanding on the less tangible, less visible aspects on the topic(s). I hope not to offend anyone's intelligence by not forming any particular opinions, critiques, conclusions, but instead offer the first thing that comes to mind... in the spirit of Maui, the innovator & inquirer.

Here are some bullet point ponderings when asked to consider the future of AI:

- Is authentic-intelligence a possible future term for that which is a natural extension & reflection of human curiosity and invention; or perhaps alliance, or affinity or animated or some "A" word that has a relational quality

- What is AI's cosmology? Or, what is the creation story we can create?

- Aside from our pedestrian & physical dependence on AI, how do we cultivate a multi-dimensional & sensual relationship to AI?

- Can we infuse a beloved tree with the technology to 'tell' us how it feels?

- In the same way that we pray to the rain cloud to disburse or collect, or to call up the fire of the volcano, could AI enhance how we communicate with elemental phenomenon & the energetic universe?

- Are there AI applications that can help us monitor how the microbiome of forests or coral reef communities are doing when we're sleeping?

- About sleep time & the super conscious & subconscious & the subliminal—when we're in states of ecstasy, meditative, trance, theta or delta states or experiences—how could we engage with AI to enhance or record inner states for quick reflection/feedback?

- Art, dance, music, poetry, kaʻao (myth) creation...I don't know what the question is besides the fact that these are necessary intelligences/processes that exercise underdeveloped parts of ourselves

I think that's it for now. Well, not really, but I'm sure we'll get to the bones of the discussion this weekend. Aloha to us all, love, Kekuhi.

How do we Indigenously Interact with AI?

Kekuhi Kealiikanakaoleohaililani

May 29th, 2019

Okay, aloha my name is Kekuhi Kealiikanakaoleohaililani. I am from Hilo from the island of Hawaii, so that's Southeast of here. My assumption is that artificial intelligence is just an extension of the human curiosity. And that I engage it every day and so do my children.

And then if that's the case then I assume that I have to create a relationship to it. That's the kind of mindset I came here with, cause I had to get a grip on something. I'm just super curious,is just how I entered this space. And if there is a challenge to bridge Hawaii life ways and some other new component of life.

My instinct is to start building the bridge. We started talking about, in the first hour, where we come from, and who we are in that community. And then I think we made the distinction about what is *not* Indigenous about how we interact with AI, and what *is* Indigenous about that.

And then I think as we became a little bit more comfortable with each other, we began to be okay with talking about, okay then if what we're looking towards is some ... An Indigenous way of having a relationship with AI, then I think we have to be okay with talking about some of our shared values. And I sort of think that's where we are right now. The Hawaii people are thinking through that and the ... The Aboriginal peoples thinking through that and Māori peoples are thinking through that. And I didn't know if we've gotten anywhere besides, the ... I think the big progression is that we're creating a new network.

Which to me is not much different from AI the interface. Let me just talk about some of the things that I've learned here, in the collective. I've learned that we all have the value of sort of inseparability with the elemental-scape. And the other thing I've learned is that we've all inherited particular cosmologies, that then sort of frame our relationship to that landscape. And the seascape and the skyscape, including the dream scape. So, if we could begin to approach AI through that story, give it a name—everything that's meaningful to us has either a name or a title or it's named a major element in the landscape, you know—and create its cosmology, because then I think our relationship with the AI structure, no matter what kind it is, we can claim as almost familial.

And I think in that way we can begin to build an Aboriginal consciousness towards our relationship with AI. And then all that requires, then, is assigning names to the parts. Like what's the name of the mineral that we begin to use to construct the actual thing? The board, the interface; what's the name of the electricity that we have to infuse into that, to the material thing?

What's the name of the silica? That when all of these parts put together creates this new sort of extension of ourselves. I don't think it's any different from having created a canoe or a net or a dream weaver or a tattoo for that matter. I think we're in a good space. I think people know enough about themselves and their place that we can come to that. I don't know if two days is enough for that conversation. But here you go, we began this symposium with an introduction that included the regular things who I am, where I come from, what is my culture, what is my tribe.

So, the reality is, is there an Indigenous world? And is there a colonized world? Or are we even permeable to the fact that as soon as you decolonize sovereignty in your own mind, there's no doubt that you can influence your family and your community and it may not be your family or community nearby you. I mean our stories aren't any different from Star Wars; it's about the hero who has to leave his community, comes out of his community, because there's only one way of thinking there.

Moves out into not just another island or another continent. He moves to another place in the universe, has his journeys, and is able to reintegrate. Now, I'm sure you have stories like that from your space— we have tons. Odysseus is another very cool example of how the human spirit is able to shift; we have to evolve. Traditions didn't become traditions because they were static. Our stories are continually changing. You cannot tell me that your grandmother told the story exactly as she heard it from your great grandparents, it's impossible. It's impossible because it's filtering through another body. There we are: we recreate the story, and if we can recreate the story, then we can do it in our own spirit. It's that central piece that I ... That's where I like to live. 10 years ago it was difficult to live there, it was challenging, and now it's the norm. I think coherence consumes incoherence. I think we have the power—as long as we maintain our relationships with the elemental world and ourself—I think we have the power to consume incoherence around us.

And we just have to stop thinking that, just stop thinking that we're only colonized, and decide who we are. And then take over the world!

How AI alters and enhances our understanding of reality

Megan Kelleher

June 25th 2019

Hi. Okay. So my name is Megan Kelleher, I'm a Barada and Gabalbara woman from Central Queensland in Australia. And so I came to be a part of this workshop through a LinkedIn connection with Angie Abdilla, and I was invited to participate because there's kind of some synergies between this work and the work that I'm doing in my PHD, looking at Indigenous knowledge systems and the blockchain. So my PHD, as I mentioned, is looking at the synergies or the conflicts between Indigenous Knowledge systems and second wave automation, artificial intelligence, blockchain and these kinds of technologies where automation is occurring. So it's really grounded essentially in Indigenous protocols and how, or whether, they can inform the design of artificial intelligence or the design of these automated protocols, these automated systems.

So I actually find AI extremely interesting because it's teaching me a lot about how, within Indigenous Knowledge systems, we're not at the centre of the universe. So I'm just finding it really interesting to learn how AI is kind of teaching me about my own culture. I'm excited by exploring what AI can do and how there are actually some different ways that cognition occurs culturally. So different cultures have different cognition processes, and so I'm interested in what AI does to time and space and how it kind of alters and enhances our understanding of reality. I'm also concerned about what it can do and what the risks might be because it's so huge and it's mysterious and it reaches into places that we don't know it's reaching a lot of the time.

And I'm concerned because do we have a choice to participate in it? And so these workshops have given me some hope, I guess, that we can influence it in an ideal world. If it does become as powerful as people are saying that it can be, I would hope that it can empower Aboriginal peoples. I hope that it can help us to understand our genius. I hope that it can help us to understand that we were always, that we always had this genius in our old ways and kind of lead us back to that place where we were before. I hope that the world listens. I hope that the people who are designing AI and using AI's and implementing AI's think really seriously about what it is that they're doing.

I hope that they get an understanding that theirs is not the only way. Our ways are valuable and important. They kept us alive. They kept the earth alive. They kept the earth healthy for thousands of generations forever into the past. So I hope that these workshops can provide some knowledge that, and I'm certain that they will. We have: we've come up with stuff that's really valuable. I just hope that people take it seriously and they don't just kind of write it off and think that's a bunch of Black fellas getting into a room and playing imaginary games. It's really important what we've done.

Our thought experiments, they will lead us somewhere if people take it seriously. You know, we've got massive fires in Tasmania. We've got massive fish kills happening in the Murray. We've got droughts happening in Queensland. We've got skeletal cattle on the front covers of newspapers. I kind of think maybe that should send some signals to people in Australia that—and not just in Australia—that's not just happening in Australia. I feel as though there might be a bit of a shift, I see little slivers of hope.

I read a story about a couple who handed back half of their property in Tasmania to an Aboriginal land council, because they believe that they can look after it and manage it better. I think that it just shows that they actually do understand and they care for the land and they want it to go on. So I think there is a little bit of a shift, however you've still got politicians in the northern territory signing off on massive fracking deals, in the face of Larrakia elders just flat out saying, no, it's not safe. We've got pipelines running through Queensland to offshore gas shipping terminals that are stirring up the reef. We've still got all of these environmental catastrophes and in some ways there is a shift, but it's far too slow. And you know, as much as AI is a really exciting area to explore the technology that it requires in its current stage, the materials that are required to support the technologies are not sustainable. So we need to think about how, if we could program an AI that can tell us: "build me with this."

This has been amazing; just coming together with all of these really thoughtful, humble, powerful, Indigenous peoples from around the world has been really inspiring and humbling. And two days has just not been long enough and I really want to be involved as the project goes forward, but I guess our message to the world, to the designers of AI and similar technologies is to be humble and remember that humans are not the centre of the universe.

Looking back to the future of AI

Maroussia Lévesque

Jan 31st 2019

In short, it looks like the past—unless we do something about it.

First, a definition. AI is an umbrella term that means different things to different people. My work focuses on machine and deep learning, because I think those are the technologies most conducive to a paradigm shift. I'll spare you the platitudes about AI's potential transformative effects, but it is worth noting that deep learning, especially in its unstructured form, can detect patterns in large data sets in a way humans can't. I'll let my comp sci colleagues unpack—or debate—this assertion.

Back to my point about the past:

- Machine and deep learning systems feed on existing data. Unchecked, they tend to reproduce and amplify existing bias. The most concerning examples sit in the criminal justice system, from predictive policing to bail determinations. Note that the latter uses a crude statistical analysis rather than complex deep learning system, but the argument stands: considering 'criminality' factors without a critical understanding of the racial and socio-economic constructs biasing the data perpetuates inequality.

- Computer science has a major white guy problem. It's important to acknowledge laudable initiatives to organize POC, non-binary and other folks, but generally AI is still designed by people who are the norm. A case in point is the lower accuracy of facial recognition systems on black and brown faces, [1] especially women's. Similarly, might a diverse team prevented the gorilla mishap? [2] To note: the company simply deleted the gorilla search results [3] rather than address the problem. There's an interesting tangential discussion about when (in)visibility is power, depending on whether AI is used in repressive contexts or to provide services. Spoiler alert: marginalized communities are overrepresented in law enforcement datasets due to over-policing. If we want AI to stop replaying the same scenario, it's time to flip the script and get a diversity of people involved upstream. Caveat: I'm also conscious/weary of the limitations of positionality, i.e. demanding that the token representative of XYZ bear the burden of defending a whole community. I think it's everyone's job, particularly those who are more privileged—a burden of proof of sorts.

- If systems are imposed top-down, marginalized/disenfranchised communities will continue to be the testbeds for oppressive practices. See Virginia Eubank's excellent case studies [4] in the US context. More broadly, AI meshes with surveillance practices in a way that challenges both domestic and international protections on privacy.

Who's Doing What

The private sector drives AI development. While some companies have called for hard regulations or international treaties, [5] the overwhelming majority lobby for soft ethical standards. Some see corporate social responsibility as a form of ethics-washing. [6] Compromises might be regulatory sandboxes, and technical standards. [Disclosure: I'm part of the IEEE standard on algorithmic bias.] [7]

Governments are also grappling with this new reality. AI-facilitated election meddling was a wake up call for many. How should nations leverage AI's economic potential, while respecting their human rights engagements? A fair criticism would be that (a) most don't and (b) human rights are a Western construct

---

[1] *Gender Shades,* <gendershades.org>.

[2] "Google apologizes after app mistakenly labels Black people 'gorillas,'" *CBC News,* July 3, 2015 <cbc.ca/news/trending/google-photos-black-people-gorillas-1.3135754>.

[3] Tom Simonite, "When it comes to gorillas, Google Photos remains blind," *Wired,* January 11, 2018 <wired.com/story/when-it-comes-to-gorillas-google-photos-remains-blind>.

[4] Virginia Eubanks, *Automating inequality: How high-tech tools profile, police, and punish the poor,* New York: St. Martin's Press, 2018.

[5] Google, Perspectives on issues in AI governance, *Google AI,* January 2019 <ai.google/perspectives-on-issues-in-AI-governance>.

[6] Ben Wagner, "Ethics as an escape from regulation: from ethics-washing to ethics-shopping?," The Privacy and Sustainable Computing Lab, 2018 <privacylab.at/wp-content/uploads/2018/07/Ben_Wagner_Ethics-as-an-Escape-from-Regulation_2018_BW9.pdf>.

[7] Algorithmic Bias Working Group, "P7003 - Algorithmic Bias Considerations," IEEE Standards Association, 2017 <standards.ieee.org/project/7003.html>.

further perpetuating oppression. At any rate, we've seen several nations and regional alliances lead consultations and issue AI strategies to hedge against perceived future risk and seek leadership in what some have called the new space race.

Ways Forward

What about people? I've already alluded to informal alliances of AI workers. Another thread is the #TechWontBuildIt [8] phenomenon. While it is not limited to AI projects, the movement opposes the use of technology for immoral purposes, and most of the actual technology involves AI. For example, Amazon employees denounced [9] the use of their facial recognition tech and cloud computing platform in support of state surveillance and immigration deportation. There's a longer discussion to be had about the potential and limitations of Valley engineers to make these kinds of decisions, but there is at least some evidence of wider coalition building with existing forms of activism.

One thing that troubles me very much is that these conversations are largely taking place without the people primarily impacted by these technologies. I've had the honor of getting a glimpse of the fierce work of the Stop LAPD Spying Coalition [10] based in Skid Row. LA is ground zero for predictive policing, and its affected communities have organized a formidable, smart response to tech-facilitated surveillance and data analytics. Coalition work is hard. It requires patience, compromise, and humility. The group must wait for everyone to be caught up and on board before it moves forward. But when it does, it speaks with a thousand voices.

I want to leave us on two more positive notes. First, art has the power to interrogate AI the way policy, law or computer science can't. I particularly enjoy the work of Trevor Paglen, [11] and I hope you will too. Back to the idea that AI is a social construct, it is largely shaped uniformly through Western concepts and values. From Estonian folklore [12] to Innu grammar [13] and Japan's Shinto tradition [14] , some concepts are making their way into AI discussions. I look forward to meeting you all and learning about what your perspectives might be.

[8] "#TechWontBuildIt," Twitter, <twitter.com/hashtag/TechWontBuildIt>.

[9] Kate Conger, "Amazon workers demand Jeff Bezos cancel face recognition contracts with law enforcement," *Gizmodo,* June 21, 2018 <gizmodo.com/amazon-workers-demand-jeff-bezos-cancel-face-recognitio-1827037509>.

[10] Stop LAPD Spying Coalition, <stoplapdspying.org>.

[11] Caitlin Hu, A MacArthur 'genius' unearthed the secret images that AI uses to make sense of us, *Quartz,* October 22, 2017 <qz.com/1103545/macarthur-genius-trevor-paglen-reveals-what-ai-sees-in-the-human-world/>.

[12] Nathan Heller, Estonia, the digital republic, *The New Yorker,* December 11, 2017 <newyorker.com/magazine/2017/12/18/estonia-the-digital-republic>.

[13] Karina Kesserwan, Indigenous conceptions of what is human, of what has a spirit and what doesn't, offer a different way of considering AI - and how we relate to each other, *Policy Options,* February 16, 2018 <policyoptions.irpp.org/magazines/february-2018/how-can-indigenous-knowledge-shape-our-view-of-ai/>.

[14] Takeshi Kimura, Robotics and AI in the sociology of religion: A human in imago roboticae, *Social Compass* 64(1), <doi.org/10.1177/0037768616683326>.

References

#TechWontBuildIt [Twitter hashtag]. (n.d.). *Twitter*. Retrieved November 11, 2019, from twitter.com/hashtag/TechWontBuildIt.

Algorithmic Bias Working Group. (2017). P7003 - Algorithmic Bias Considerations. *IEEE Standards Association*. Retrieved from standards.ieee.org/project/7003.html.

Conger, K. (2018, June 21). Amazon workers demand Jeff Bezos cancel face recognition contracts with law enforcement. *Gizmodo*. Retrieved from gizmodo.com/amazon-workers-demand-jeff-bezos-cancel-face-recognitio-1827037509.

Eubanks, V. (2018). *Automating inequality: How high-tech tools profile, police, and punish the poor*. New York: St. Martin's Press.

Gender Shades. (n.d.). Retrieved November 11, 2019, from gendershades.org.

Google apologizes after app mistakenly labels Black people 'gorillas' [online article]. (2015, July 3). *CBC News*. Retrieved from cbc.ca/news/trending/google-photos-black-people-gorillas-1.3135754.

Google. (January 2019). Perspectives on issues in AI governance [PDF document]. *Google AI*. Retrieved from ai.google/perspectives-on-issues-in-AI-governance.

Heller, N. (2017, December 11). Estonia, the digital republic. *The New Yorker*. Retrieved from newyorker.com/magazine/2017/12/18/estonia-the-digital-republic.

Hu, C. (2017, October 22). A MacArthur 'genius' unearthed the secret images that AI uses to make sense of us. *Quartz*. Retrieved from qz.com/1103545/macarthur-genius-trevor-paglen-reveals-what-ai-sees-in-the-human-world.

Kesserwan, K. (2018, February 16). Indigenous conceptions of what is human, of what has a spirit and what doesn't, offer a different way of considering AI - and how we relate to each other. *Policy Options*. Retrieved from policyoptions.irpp.org/magazines/february-2018/how-can-indigenous-knowledge-shape-our-view-of-ai.

Kimura, T. (2017, January 30). Robotics and AI in the sociology of religion: A human in imago roboticae. *Social Compass* 64(1). Retrieved from doi.org/10.1177/0037768616683326.

Simonite, T. (2018, January 11). When it comes to gorillas, Google Photos remains blind. *Wired*. Retrieved from wired.com/story/when-it-comes-to-gorillas-google-photos-remains-blind.

Stop LAPD Spying Coalition. (n.d.). Retrieved November 11, 2019, from stoplapdspying.org.

Wagner, B. (2018). Ethics as an escape from regulation: From ethics-washing to ethics-shopping? *The Privacy and Sustainable Computing Lab*. Retrieved from privacylab.at/wp-content/uploads/2018/07/

Ben_Wagner_Ethics-as-an-Escape-from-Regulation_2018_BW9.pdf.

Will Indigenous ways of thinking save AI?

Keoni Mahelona

February 27, 2019

I rarely blog. [1] Not good at it. In 3rd grade I was put into the special reading class. Reading and writing was never my thing, but I always loved math and science and all disciplines derived from those fundamental subjects.

I'm attending an Indigenous AI workshop in Hawaiʻi. I initially thought this was gonna be a brown nerd meetup 😄 but it's much better than that. The point is to bring together Indigenous and some non-Indigenous doers, makers, and creators to discuss what Indigenous AI is and how it *will* play an important role in the future of AI for humanity.

It's probably best to insert my background [2] here to justify why you should even consider what I have to say on the matter. I won't do that. Those who know the work I do, which are primarily the communities I serve, know me and respect my whakaaro. [3] That's important here—community and trust. I'll try to link that in later (again I'm not a good writer)

So the question I have to answer is "what does the future look like for AI?" I'll answer this question purely based on what I know now from the work I've done over the years in science and engineering as a Kanaka Māoli.

I need to preface that I'll use machine learning and AI interchangeably. Machine learning is a tool that might lead to artificial intelligence, but I don't think that will happen. Peter Lucas Jones [4] (also attending the workshop) says it best, "Ko te AI tētahi karetao ka taea e tātou te whakakōrero me te whakakanikani. Mā te whakamahi i o tātou rarāunga me ngā kōrero tuku iho, ka tutuki ngā āhutanga o te karetao." He's basically saying AI is a puppet and we make it do what we want using our data and knowledge. **Puppet.** Until we figure out a way to do AI that isn't only data driven, I don't think we'll reach the singularity.

For me, the future for AI is looking bad. Currently the big corporates (the wealthy, the 1%, the colonizers,

---

[1] Originally published K. Mahelona (2019) Will indigenous ways of thinking save AI?, *Medium*, <medium.com/@mahelona/what-does-the-future-look-like-for-ai-1ffdff620395>.

[2] Keoni Mahelona, "Keoni Mahelona - CTO - Te Hiku Media," *LinkedIn* <linkedin.com/in/kmahelona>

[3] Search results for 'keoni mahelona', *Te Hiku Media* <tehiku.nz/search?q=keoni%20mahelona>.

[4] peterlucasjones, *Twitter*, <twitter.com/peterlucasjones>.

etc.) are leading the way in AI. The current technology trends show that you need vast amounts of data and huge computational power to achieve anything close to 'AI.' The scales at which AI works are financially unreachable by most people, and I find this terribly frightening—corporates have more power in AI than sovereign nations (that's nothing new in colonial history—profits drove much of colonization including the overthrow of the Hawaiian Kingdom with the illegal aid of the U.S. Military).

Having said that, a small non-profit, Te Hiku Media,[5] is able to deploy its own speech recognition software[6] in the cloud thanks to services like AWS and open source projects like Mozilla's DeepSpeech.[7] In this case, machine learning is just another tool to help us do what we need to.

The difference between Te Hiku Media's 'A'' and Google's 'A'' is that ours is created from our Indigenous language—our data. We collected this data. We look after this data with *tikanga* (cultural practices and values). We will not allow large corporates to have access to this data and use it to exploit us (e.g. serve us ads, sell our language as a service back to us, read our cultural knowledge, etc.). This data is unique to our people, about 600k Māori, the Indigenous people of Aotearoa. We were able to collect this data because the community that shared it with us trusts us. We've worked with the community and for the community for the last 30 years. Our data is what makes us unique. It is our own 'AI,' the puppet we've created to help us achieve our goals and aspirations as a people revitalizing our reo.

This is where data sovereignty—privacy and guardianship over individual data and the data of groups of people—is critically important. If we can maintain that sovereignty, we can prevent the 1% from further colonizing us. But I see the opposite happening. Global corporates like Lionbridge are soliciting Indigenous people to sell them their language—they'll pay you USD$45 for 1 hour of your time. They clearly have customers in mind as they're a globalization and localization company. You see companies like Duolingo and Drops being given our languages for the sake of revitalisation and promotion. And while these companies might be good at heart, they make a profit from selling language services. Do those profits make their way back to our communities from which the language data was taken? Or should we be thanking them as the saviours of our people and they can have our data for free... what ever happened to all our land? Of course the biggest insult comes from DNA companies like Ancestry.com. **YOU PAY THEM** to **GIVE THEM YOUR GENETIC DATA**, and they have the right to use it as they deem fit. Read the terms and conditions whānau! AI is very much about our data and our knowledge.

Don't get me wrong. I know society as a whole could benefit when we share genetic data, when we open source knowledge, and when we put data in the public domain. But in a world with so much inequality, racism, genocide, the list goes on and on, clearly only the wealthy are to benefit from these 'public' goods and services.

[5] *Te Hiku Media,* < tehiku.nz/>.

[6] *kōreromāori.io,* <koreromaori.io>.

[7] "Mozilla/DeepSpeech: A TensorFlow implementation of Baidu's DeepSpeech architecture," *GitHub,* <github.com/mozilla/DeepSpeech>.

I wish AI could change the balance of power, but I can't imagine that happening anytime soon. It's possible that a technological revolution could do the trick. If/when quantum computers (or some computationally equivalent tech) exist at the consumer level, that could give the 99% similar power to the 1%. But history dictates that the technology itself isn't enough to 'do good.' We need laws and ethics around the technology that guides its use for the benefit of all of humanity (and the planet) and not just the wealthy, pale, stale, and males. Chief Sitting Bull made such a keen observation in the 19th century that still stands today, "the white man knows how to make everything, but he does not know how to distribute it." He said this on reflection of the white man's neglect for their poor. With all the Western wealth and technologies in 2019, we still can't solve such a basic problem as poverty.

Western science is only just recognizing how Indigenous knowledge can help our planet, especially in the face of environmental destruction and climate change. I believe how Indigenous people look after their data and knowledge could also help form a framework for AI that works in the best interest of everything contained within our solar system. We personified land and water not because we were hedonistic, demigod worshipers, but because these personifications allowed us to maintain a level of respect and responsibility toward our environments.

I think AI will reaffirm Indigenous knowledge especially around the fringes of science. For example, how are humans affected by the moon, *māramataka*? There's a huge body of traditional knowledge around that and while western science might call this new age mumble jumble (thanks hippies!), the data I've observed—people around me have cycles of behavior aligning with the lunar cycle—is enough for me to say, hey, how could we measure these behaviors and use them to predict patterns? Machine learning could help us understand from a western perspective some of what we know already know in an Indigenous context.

For an AI to not be a puppet, I think it needs to be able to do something as basic as caring for the poor *without* being forced to do so. It's one thing to force people to pay taxes and another for people to fundamentally understand the value and joy in paying taxes in a civilised society. I live in New Zealand. I do enjoy paying taxes because I know it means I get free health care and it helps with the conservation and protection of New Zealand ecosystems. I would not enjoy paying taxes in the U.S. because it funds genocide, colonisation, and the wealthy.

But what creates that difference between being forced to do good and having joy in doing good? I suppose that's nurture. How we grow up, the people we are surrounded by, and the communities we belong to all come together to shape *why* we do the things we do. We're a reflection of our environment, or rather the data we're exposed to determines whether we want to do the things we do or whether we're forced to do the things we do. If this is the case, I do not trust the Big Five to build AI, and I do not trust countries like the US and China to build AI. I'd really only trust an AI coming from my own people and the communities of which I am apart. Huh, I'd say the same is true for humans I trust.

References

Jones, P. L. peterlucasjones. (n.d.). *Twitter*. Retrieved November 11, 2019, from twitter.com/peterlucasjones.

kōreromāori.io. (n.d.). Retrieved November 11, 2019, from koreromaori.io.

Mahelona, K. (n.d.). Keoni Mahelona - CTO - Te Hiku Media [LinkedIn profile]. Retrieved November 11, 2019, from linkedin.com/in/kmahelona.

Mahelona, K. (2019, February 27). Will Indigenous ways of thinking save AI? *Medium*. Retrieved from medium.com/@mahelona/what-does-the-future-look-like-for-ai-1ffdff620395.

Mozilla/DeepSpeech: A TensorFlow implementation of Baidu's DeepSpeech architecture [repository]. (n.d.). *GitHub*. Retrieved November 11, 2019, from github.com/mozilla/DeepSpeech.

Search results for 'keoni mahelona' [webpage]. (n.d.). *Te Hiku Media*. Retrieved November 11, 2019, from tehiku.nz/search?q=keoni%20mahelona.

Te Hiku Media. (n.d.). Retrieved November 11, 2019, from tehiku.nz.

Caleb Moses on the Bleeding Edge

Caleb Moses

June 12, 2019

My name is Caleb Moses. I'm a data scientist from New Zealand, based in Auckland. I'm working for Dragonfly Data Science, and we are working with Te Hiku Media on an exciting body language technology project. So we built the first speech-to-text algorithm in Te Reo Māori. So where you can speak Te Reo, the Māori language, to your computer and then it will be able to transcribe what you're saying in real time. My relationship to AI is that I like to build them.

At university, I did mathematics and when I graduated, I was looking, you know, what are the interesting maths jobs that I could go and apply for. That's how I learned about data science, how I learned about machine learning. I spent about a year, well, no. I spent about a year studying on my own, pretty hardcore, and then another year trying to apply it in my work and, eventually, found myself working at Dragonfly with Te Hiku.

So, personally, I'm more interested in using AI as a tool to create things. Basically, what you do if you are kind of interested in AI and stuff is you find the people who are on the bleeding edge, and you follow them. You follow them on Twitter. You see what work they're doing. You see the stuff that's coming out of the big labs, DeepMind and Google Brain and Uber and all of the stuff that they're doing, Facebook. Then you try to figure out how you can take those technologies, and then use them on your scale because one of the big problems is not just access to know-how.

Because, generally speaking, at least for someone like me who has a university degree and that sort of thing, there's a lot of resources available online where you can go and learn how to put these things together yourself. I know that for Indigenous communities where university degrees are in short supply, that's not necessarily what could be considered easy access. But at least for me, I've been able to kind of find stuff online, find blog posts, read them, figure out how to put them together, how to run the models myself. I've been able to do that.

But one of the really big gaps between us and Facebook is just computational power. They have so many more computers than we could ... We can scarcely imagine how many computers they have. I remember finding out a few years ago that Google ...that they had built this new kind of hardware to do AI models real fast. I managed to find the source that said that their models ... like they have so much computing power that they can run, like, object recognition across all of Google Maps, like all of the street view for the entire world, they can do it in about two days, which is like, yeah. Yeah, there's no way that a person could do that. There's no way that a university could do that. Yeah, it's totally insane. I've been excited being here at the conference getting to talk with people who have access to more Indigenous data than what I've been able to find so far. Te Hiku themselves have probably, so far as I know, the best collection of at least Māori audio, but probably also Māori text now that we've assembled our language corpus, and I'm definitely looking forward to doing some interesting work with that.

Just a few weeks ago, I was working on a model that basically generated Māori language text. You just feed it all corpus and then it learns how to make new stuff. It went pretty well, but I think it could do a lot better and, yeah, just more work. More work needs to happen in this area. I think that that's another thing that, at least as Indigenous people, we could really kind of leverage that knowledge that we have about where we come from to create new things. And not just new things, but new things that only we can make, or at least that only we should make. So, that's what I'm excited about.

My dream, and I say dream sort of on purpose, I want to see an Indigenous AI research lab that creates things that are Indigenous, yes, but also things that are on the bleeding edge with everyone else. That's what I want to see, so I want to see us making our own image recognition algorithms, and our own AIs that play chess better than humans and all of that sort of stuff. But also using that knowledge to create new things: new ways of interacting with our culture, like building new tech, the stuff that Te Hiku are doing now, voice recognition in Te Reo Māori. We could create our own virtual assistance. We could bake them into video games. People could play video games where they have to say a spell in Te Reo Māori in order for it to work. My idea it would be us kind of creating new technologies that are just out there with the best of them. That's what I think. That's what I think we can do.

What does the future look like for AI?

'Ōiwi Parker Jones

February 28, 2019

What is Indigenous AI?

I suppose that answering this question will be part of our task at the workshop.

My first thought on the topic was to frame it in terms of AI *by* Indigenous communities and for Indigenous communities. But I have also, more recently, been considering a third way: *AI in dialogue with Indigenous communities.*

I would propose the following working definitions for the three views:

(1) AI *by* an Indigenous community is AI that is produced by one or more members of an Indigenous community.

(2) AI *for* an Indigenous community is AI that addresses the needs of one or more Indigenous communities.

(3) AI *with* an Indigenous community is AI that is in dialogue with one or more Indigenous communities.

Here (1) is intended to denote anything produced *by* a member of an Indigenous community, no matter what. So if a member of an Indigenous community worked on any random topic in machine learning, then, by definition, it would be 'Indigenous AI'. To me this misses the point. As an Indigenous person who works on AI, I appreciate the sentiment. But if I invented a new kind of LSTM module, should that module be considered 'Indigenous AI'? We could end up with an incoherent subset of AI research that we call 'Indigenous AI' simply because Indigenous people worked on those things.

Definition (2) focuses on the content of the research, rather than on the identity of the researcher. AI *for* an Indigenous community might include some of my own work on Hawaiian NLP. Should any research that touches on topics relevant to an Indigenous community be considered 'Indigenous AI'? One limitation of (2) is that it does not give agency to our Indigenous communities over what counts as 'Indigenous AI'. Any company might, for example, develop an application for one of our languages, or for any part of our culture, and then market it as 'Indigenous AI'. Is that the space that we want to create around this term?

Definition (3) is meant to maximise the pros and minimise the cons of (1) and (2). 'In dialogue with' is meant to express the idea that the AI is being actively engaged with by an Indigenous community.

One reason to engage with AI research might be that it is being performed by someone who is already a member of the community, as in (1). Another reason is that the AI research bears on topics that are important to the community, as in (2). But definition (3) leaves the choice about what counts as 'Indigenous AI' up to our communities, so that that it should be impossible to hijack the term without buy-in from at least one of our communities.

This, I would suggest, is one way to frame what we will be doing at the workshop: entering into dialogue between research on AI and our Indigenous communities.

From this perspective, then, what does the future of Indigenous AI look like? This question has been posed by the workshop organisers. If I could suggest a few relevant topics, they would include: intellectual property, fairness, and data-efficiency. I hope that we will get to talk more about these things at the workshop. However, if the big idea is to create a community of ideas, then I also look forward to finding out what 'Indigenous AI' means together.

I also hope that we might continue to broaden our conversation to include more non-Indigenous AI researchers, with the intention of producing as active an ecosystem of ideas together as we can.

What does the future look like for AI?

Caroline Running Wolf

February 18, 2019

As a preschooler I was fascinated by my friend's parents, who have been researching and trying to develop an artificial intelligence for a large company since the 1960's. Whenever I checked in with them, every decade or so, they laughed it off and confided in me that artificial "intelligence" still had a long way to go to fill the shoes of that label.

Today we have achieved a certain level of (almost) artificial intelligence—for clearly delineated, specific tasks. Much of this is still based on computational pattern recognition through large amounts of data. Machines still can't learn and infer context like humans can. But humans are the ones programming these machines—and it shows.

On a regular basis reports surface about AI powered software with racial or gender bias. Earlier this month a Twitter user posted a screenshot of a suggested correction by Grammarly, an online grammar and contextual spell checking platform. Grammarly had an issue with an "unusual word pair" and suggested to combine the noun "girl" with an adjective other than "successful," positing that synonyms like "lucky" or "happy" might be more fitting. Facial recognition software jumps from a 1% error margin for light-skinned males to over 35% for dark-skinned women. Despite the obvious bias in current AI systems, Joy Buolamwini, founder of the Algorithmic Justice League, concludes her February 7, 2019 *Time* article on a hopeful note:

"I am optimistic that there is still time to shift towards building ethical and inclusive AI systems that respect our human dignity and rights. By working to reduce the exclusion overhead and enabling marginalized communities to engage in the development and governance of AI, we can work toward creating systems that embrace full spectrum inclusion. In addition to lawmakers, technologists, and researchers, this journey will require storytellers who embrace the search for truth through art and science. Storytelling has the power to shift perspectives, galvanize change, alter damaging patterns, and reaffirm to others that their experiences matter. That's why art can explore the emotional, societal, and historical connections of algorithmic bias in ways academic papers and statistics cannot. And as long as stories ground our aspirations, challenge harmful assumptions, and ignite change, I remain hopeful." [1]

I agree with Joy Buolamwini. Despite currently manifested biases and limitations, the future for AI is still malleable. Our workshop is not a day too early!

Today's implementations of AI are already very promising. Personally, I am excited about the possibilities of AI, especially what speech recognition, Natural Language Processing (NLP) and chat bots offer for the revitalization of endangered Indigenous languages. This is the field that I am passionate about and I am willing to recruit the help of any technology available for this goal. I realize that the amount of data needed for NLP to generate speech and interactions for Indigenous languages are a major hurdle—but just imagine the possibilities!

Some endangered Indigenous languages have only a handful of fluent speakers left. These speakers are elderly. Our time with them is limited and we have to use it wisely. We shouldn't waste their energy and knowledge by making them teach language beginners or having them translate individual words for a dictionary. Technology can assist with these simple tasks. In the future, home assistants could be programmed to recognize and respond in Indigenous languages, allowing language learners to apply and practice their language skills. Real-time translation could translate websites and social media as well as dub TV shows and movies. We could interact with video game characters in our Indigenous language, engaging in human-like conversations. With the help of current and future AI technologies we can build language tools that expand our everyday usage of Indigenous languages.

No technology can replace humans and true human interaction but just like other technologies that came before it, artificial intelligence can change our lives. My hope is that AI will also have a major effect on the reclamation of our Indigenous languages.

References

[1] Joy Buolamwini, "Artificial intelligence has a problem with gender and racial bias. Here's how to solve it," *Time*, February 7, 2019 <time.com/5520558/artificial-intelligence-racial-gender-bias>.

Buolamwini, J. (2019, February 7). Artificial intelligence has a problem with gender and racial bias. Here's how to solve it. *Time*. Retrieved from time.com/5520558/artificial-intelligence-racial-gender-bias.

What does the future look like for AI?

Michael Running Wolf

February 21, 2019

The future is the continuing proliferation and accessibility of Machine Learning (ML). Though the fundamental math and technology has not changed, the access and relative ease to create advanced AI systems has. A mere generation ago custom built supercomputers, and millions of dollars of investment, was the minimal entry fee to use ML. Now, in addition to the advent of the Open Source Software (OSS) movement, ML is consumer grade. One could build a reasonable ML computer with top of the line software tooling for less than $1,000! Even that is not strictly necessary, all you need is a web browser to access cloud computing. One could, for a fee, deploy a supercomputer cluster within minutes. For Indigenous nations, this access is at once an opportunity and risk.

The AI tooling to suppress Native activists, protecting sacred lands, is easily purchased by antagonistic special interests. One not need be a well financed national state, small agencies can easily license facial recognition software to monitor 'radical environmentalists' protecting their sacred lands from exploitation. Advanced facial recognition turns any phone into a potential spy while social media photo platforms are susceptible to analysis. Though our privacy is at risk, the benefits outway the risk.

Every internet user is a few minutes away from deploying their very own ML infrastructure and a wealth of research. TensorFlow, the most popular ML framework for instance, is freely available and gives community researchers access to millions of dollars of research development investment. We are limited only by time and skill.

Initially, a tribe's community researchers could collect the decades of anthropological and linguistic research collected in mountainous digital archives. A researcher can expect to barely scratch the surface of this knowledge if they diligently read every word. However, with advanced text analysis one can quickly mine the knowledge to rediscover lost insights into their own tribe. These insights can then form the building blocks for advanced cultural and linguistic revitalization tooling.

For example, one could textmine the Hawaiian news archive, the Papakilo Database, and build a statistical language corpus. With this corpus one could train recognition and generative ML systems, i.e. a way of validating proper Hawaiian grammar while also creating a mechanism to generate new sentences. With these tools in hand one can create a Hawaiian chatbot! With phonemes and audio recognition you are inches away from creating an Indigenous Voice AI similar to Apple Siri or Google Assistant. Imagine Virtual Reality worlds populated by intelligent Hawaiian language speakers wanting nothing more than to teach you a new language. Everyone needs an infinitely patient Indigenous personal language teacher.

Despite the risk, ML offers opportunity for Indigenous communities. In fact we have little choice, Machine Learning will be leveraged against us or by us.

An Urban Mohawk Woman Who Loves Her Cyberpunk Avatar Envisions The Future Of AI

Skawennati

2019-02-26

Sken:nen,

While I wouldn't call myself a Trekkie, I am a Star Trek fan. My favourite series was *The Next Generation*. I love how Star Trek portrays the future: filled with space-faring human/alien half-breeds and higher intelligences, yet governed by the Prime Directive, which privileged knowledge exchange over slavery or other resource extraction. Mr. Data was the show's portrayal of Artificial Intelligence. Housed in a humanoid cyborg body (some would say he's an android—but not I), he could do many things that only the computers in previous iterations of the show could do, such as scan a planet for life forms.

The majority of my ideas about AI come from fictional books and movies like *Neuromancer* and *The Terminator*. Most recently, I've become fascinated by the portrayal of the AI from the Netflix series *Travellers*. *The Director*, as It is known, is revered by the people of the future as if It were a god. It (and it is emphatically an "It") only shows up in computer code (although sometimes, if absolutely necessary, It can inhabit a child's body). Through the omni-present surveillance devices of contemporary life, as well as the time travelling agents sent to present-day Earth, The Director is able to see all. Its job is to figure out what events in the past should be altered or avoided so that the Earth does not become the barren wasteland it is in the future where It is from.

The AIs of today are much less exciting than the AIs of fiction. As Nick Heath of ZDNet says, in an informative article called "What is AI? Everything you need to know about Artificial Intelligence": [1] "AI is ubiquitous today, used to recommend what you should buy next online, to understand what you say to

---

[1] Nick Heath, "What is AI? Everything you need to know about artificial intelligence," *ZDNet*, February 12, 2018 <zdnet.com/article/what-is-ai-everything-you-need-to-know-about-artificial-intelligence/>.

virtual assistants, ... to recognise who and what is in a photo, to spot spam, or detect credit card fraud."

I am happy that Gmail's AI filters out my spam, and my bank sends me a new card when some thief gets their hands on my number. For these AIs I am thankful. I do sometimes wonder, however, what we might be missing out on. It seems like the AI-makers think that it's a small price to pay if one real email gets lost in the spam. But what if that is the golden email?

I recently met an artist who is using AI to create paintings (reading their blog reminds me of how little I know about real-life AI and machine learning. Sorry folks.). They are using a machine-learning algorithm with multiple discriminators to generate unique works of art. What I understand from that, as well as from a conversation I had with them, is that the AI is composing the image, selecting the colours, determining the style, and ensuring technical merit. "But that's all the fun stuff!" I said to them in dismay. And why in the world do they want to put artists out of work?

You asked us what the future looks like for AI.

For one thing, I don't think AIs will look human, the way the AI child looks in the movie *AI*. I think we are smart enough to avoid that folly. I think they'll probably become more like avatars that we each customize, like a visual Samantha from *Her*.

Also, I don't think AIs will want to be human. I read a great quote (that I forgot to cite) that says that "humankind has a massive ego thinking that we are the center of the universe and everything around us must desire us in some capacity,."

Which brings me to this workshop.

I am excited by the idea that we are engaging with AI on our terms, as Indigenous people. I am excited that a platform is being built such that other, non-Indigenous folk might listen to what we have to say on this topic.

The strength of an AI—its very raison d'être—is that is can solve complex problems. Perhaps it can solve the problem of social injustice. Maybe it can figure out how to bring about a non-violent revolution.

I have been reading about the history of the confederation of the Haudenosaunee. The three tenets of the Great Law of Peace, which is our constitution, were peace, unity and the good mind. My ancestors had in place a complex system of consensus in order to come to decisions. I wonder if we could feed that info to the AI?

At the very least, we need to program the AI with the Thanksgiving Address, the oral tradition that reminds us of the familial relationships between the earth, water, sky and all the things living there. Most of us Indigenous folk have a similar teaching or ceremony. That ancient message is very similar to Star Trek's message. As Kyle Sullivan and Katie Boyer of Trekpertise [2] say, it is meant to "remind us to show respect and reverence for all life, and forms of intelligence, whether natural or artificial."

References

Evans, C. (2016, November 4). Artificial intelligence in Star Trek. *Redshirts Always Die*. Retrieved from redshirtsalwaysdie.com/2016/11/04/artificial-intelligence-star-trek.

Heath, N. (2018, February 12). What is AI? Everything you need to know about artificial intelligence. *ZDNet*. Retrieved from zdnet.com/article/what-is-ai-everything-you-need-to-know-about-artificial-intelligence.

---

2   Charles Evans, "Artificial intelligence in Star Trek," *Redshirts Always Die*, November 4, 2016 <redshirtsalwaysdie.com/2016/11/04/artificial-intelligence-star-trek/>.

6.2

Indigenous Protocol and AI Reading List

The following is a list of resources that workshop participants drew upon in their discussions. It is not comprehensive, and, in fact, is somewhat idiosyncratic.

Indigenous Knowledge + AI and Digital/Computational Technology

Abdilla, A. (2018). Beyond imperial tools: Future-proofing technology through Indigenous governance and traditional knowledge systems. In Harle, J., Abdilla, A., & Newman, A. (Eds.), *Decolonizing the digital: technology as cultural practice* (pp. 67–81). Sydney, AU: Tactical Space Lab.

Abdilla, A., & Finch, R. (2016). Indigenous knowledge systems and pattern thinking: An expanded analysis of the first Indigenous robotics prototype workshop. *The Fibreculture Journal: Digital Media + Networks + Transdisciplinary Critique*, (28). Retrieved from twentyeight.fibreculturejournal.org.

Bourgeois-Doyle, D. (209) "Two-Eyed AI: A Reflection on Artificial Intelligence." The Canadian Commission for UNESCO's IdeaLab.

Catlin, D., Smith, J. L., & Morrison, K. (2012). Using educational robots as tools of cultural expression: A report on projects with Indigenous communities. In Obdržálek, D. (Ed.), *RiE 2012: 3rd international conference on robotics in education - Conference proceedings* (pp. 73-79).

Crembil, G., & Gaetano Adi, P. (2017). Mestizo robotics. *Leonardo, 50*(2), 132–137. doi.org/10.1162/LEON_a_01150.

Gasparotto, M. (2016). *Digital colonization and virtual Indigeneity: Indigenous knowledge and algorithm bias.* Manuscript for the Annual Conference of the Seminar on the Acquisition of Latin American Library Materials

Kwaymullina, A. (2017). Reflecting on Indigenous worlds, Indigenous futurisms and artificial intelligence. *Mother of Invention.* Retrieved from motherofinvention.twelfthplanetpress.com/2017/09/16/reflecting-on-indigenous-worlds-indigenous-futurisms-and-artificial-intelligence.

Kesserwan, K. (2018). How can indigenous knowledge shape our view of AI? *Policy Options.* Retrieved from policyoptions.irpp.org/magazines/february-2018/how-can-indigenous-knowledge-shape-our-view-of-ai.

Lewis, J. E. (2019). An orderly assemblage of biases: Troubling the monocultural stack. In Schweitzer, I. (Ed.), *Afterlives of Indigenous archives* (pp. 219–31). Lebanon, MA: New England Press.

___. (2014). A better dance and better prayers: Systems, structures, and the future imaginary in Aboriginal new media. In S. Loft & K. Swanson (Eds.), *Coded territories: Tracing indigenous pathways in new media art* (pp. 48–77). Calgary, CA: University of Calgary Press.

Lewis, J. E., Arista, N., Pechawis, A., and Kite, S. (2018). Making kin with the machines. *Journal of Design and Science.*

Lozano-Hemmer, R. (1996). FLOATING TROUT SPACE - native art in cyberspace. *Telepolis.* Retrieved from heise.de/-3441019.

Martínez, Christopher. (2015). Tecno-sovereignty: An Indigenous theory and praxis of media articulated through art, technology, and learning (Doctoral dissertation). Retrieved from ProQuest Dissertations & Theses Global database. (Accession No. 3701432).

Ozichi Emuoyibofarhe, N., Segun, A., Olusegun Lala, G., & Omolola Aremu, R. (2015). A Yoruba cultural tradition repository knowledge based system. *International Journal of Emerging Trends in Science and Technology, 2*(7): 2830–41.

Phahlamohlaka, L.J., and Kroeze, J.H. (2005). Sacred space in cyberspace: An African perspective. *Journal for Semantics - Tydskrif Vir Semitistiek,* 14(2): 413–40.

Todd, L. (1996). Aboriginal narratives in cyberspace. In M.A. Moser & D. MacLeod (Eds.), *Immersed in technology: Art and virtual environments* (pp. 179–194). Cambridge, MA: MIT Press.

Indigenous Epistemology, Ontology, Cosmology and Ethics

Waters, A. (Ed.). (2004). *American Indian thought: Philosophical essays*. Oxford, GB: Blackwell Publishing.

Little Bear, L. & Heavy Head, R. (2004). A conceptual anatomy of the Blackfoot world. *ReVision*, 26(3), 31-38.

Wilson-Hokowhitu, N. (Ed.). (2019). *The past before us: Moʻokūʻauhau as methodology*. Honolulu, HI: University of Hawaii Press.

Cajete, G. (2000). Native science: *Natural laws of interdependence* (1st ed.). Santa Fe, NM: Clear Light Publishers.

de Castro, E. V. (2004). Exchanging perspectives: The transformation of objects into subjects in Amerindian ontologies. *Common Knowledge 10*(3), 463-484.

Cheney, J. (1989). Postmodern environmental ethics: Ethics of bioregional narrative. *Environmental Ethics 11*(2), 117-134.

Cheung, M. J., Gibbons, H. M., Dragunow. M., & Faull, R. L. M. (2007). Tikanga in the laboratory: Engaging safe practice. *MAI Review 1*, 1-7.

Descola, P. (2013). *Beyond nature and culture*. Chicago, IL: University Of Chicago Press.

Hernandez, N. (1999). Mokakssini: A Blackfoot theory of knowledge (Doctoral dissertation). Harvard University, Cambridge, MA.

Hester, L. & Cheney, J. (2001). Truth and Native American epistemology. *Social Epistemology, 15*(4), 319-334.

Kuwada, B. K. (2015, April 3). We live in the future. Come join us [Blog post]. Retrieved from hehiale. wordpress.com/2015/04/03/we-live-in-the-future-come-join-us.

Meyer, M. A. (2003). *Hoʻoulu: Our time of becoming - Collected early writings of Manulani Meyer*. Honolulu, HI: Short Stack Native Books.

Nakata, M., Hamacher, D., Warren, J., Byrne, A., Pagnucco, M., Harley, R., ... Bolt, R. (2014). Using modern technologies to capture and share Indigenous astronomical knowledge. *Australian Academic & Research Libraries, 45*(2), 101–110.

Posthumus, D. (2018). *All my relatives: Exploring Lakota ontology, belief, and ritual*. Lincoln, NE: University of Nebraska Press.

Rainforth, D. How Aborigines invented the idea of object-oriented ontology [Supplemental material].

Un Magazine, 10(1). Retrieved from unprojects.org.au/magazine/issues/issue-10-1/object-oriented-ontology-web-only.

Silva, N. K. (2017). *The power of the steel-tipped pen: Reconstructing native Hawaiian intellectual history.* Durham, NC: Duke University Press.

Smith, L. T. (2012). *Decolonizing methodologies: Research and Indigenous peoples.* London, GB: Zed Books Ltd.

Turner, D. (2006). *This is not a peace pipe: Towards a critical Indigenous philosophy.* Toronto, CA: University of Toronto Press.

Industrial Artificial Intelligence

AI for humanity [Webpage]. (n.d.). Retrieved from mila.quebec/en/ai-society.

Bell, G. (2018, March). *The AI revolution - Human-computer relationships in the 4th industrial age.* Presentation at the 2018 Royal Australian Air Force Air Power Conference, Canberra, AU. Video recording retrieved from youtube.com/watch?v=_zAUkhoruk8.

___. (2017, July). *Putting AI in its place: Why culture, context and country still matter.* Presentation at the AI Now 2017 Public Symposium, New York, NY. Video recording retrieved from youtube.com/watch?v=WBHG4eBeMXk.

Benkler, Y. (2019). "Don't let industry write the rules for AI." *Nature,* 569(161), 161.

Crawford, K. & Joler, V. (2018). Anatomy of an AI system. Retrieved from anatomyof.ai.

Elements of AI [Online course via Google Digital Garage]. (n.d.). Retrieved from learndigital.withgoogle.com/digitalgarage/course/elements-artificial-intelligence.

Facebook Research. Introduction to AI [Video]. (2016). Retrieved from research.fb.com/videos/introduction-to-ai.

Gebru, T. (2019, March). *Understanding the limitations of AI: When algorithms fail.* Global Women in Data Science Conference, Stanford University, CA. Video recording retrieved from youtube.com/watch?v=Q2CSVYD7pYE.

Google. Perspectives on Issues in AI Governance. (2019). Retrieved from ai.google/static/documents/perspectives-on-issues-in-ai-governance.pdf.

Introduction to machine learning [Online course via Google Developers]. (n.d.). Retrieved from developers.google.com/machine-learning/crash-course/ml-intro.

Onuoha, M., & Nucera, D. (2018). *A people's guide to AI*. Allied Media Projects. alliedmedia.org/peoples-ai.

Parker Jones, O., & Shillingford, B. (2018, December). Composing RNNs and FSTs for small data: Recovering missing characters in old Hawaiian text." Paper presented at the 32nd Conference on Neural Information Processing Systems, Montréal, CA. Retrieved from researchgate.net/publication/338047653_Composing_RNNs_and_FSTs_for_Small_Data_Recovering_Missing_Characters_in_Old_Hawaiian_Text.

Winograd, T., & Flores, F. (1987). *Understanding Computers and Cognition: A New Foundation for Design*. Boston, MA: Addison-Wesley.

3.5 Resisting Reduction Competition Winners. (2018). Retrieved from jods.mitpress.mit.edu/competitionwinners.

Algorithmic Bias & Data Sovereignty

Angwin, J., Larson, J., Mattu, S., & Kirchner, L. (2016). Machine bias. Retrieved from propublica.org/article/machine-bias-risk-assessments-in-criminal-sentencing.

Cave, S. (2017). Intelligence: A history. Retrieved from aeon.co/essays/on-the-dark-history-of-intelligence-as-domination.

Crawford, K. (2013). The hidden biases in big data. Retrieved from hbr.org/2013/04/the-hidden-biases-in-big-data.

Corbett-Davies, S., Pierson, E., Feller, A., & Goel, S. (2016). A computer program used for bail and sentencing decisions was labeled biased against Blacks. It's actually not that clear. Retrieved from washingtonpost.com/news/monkey-cage/wp/2016/10/17/can-an-algorithm-be-racist-our-analysis-is-more-cautious-than-propublicas/.

Duarte, M. E. (2017). *Network sovereignty: Building the internet across Indian Country*. Seattle, WA: University of Washington Press.

Gasparotto, M. (2016). Digital colonization and virtual Indigeneity: Indigenous knowledge and algorithm bias.

Kukutai, T., & Taylor, J. (Eds.). (2016). Indigenous data sovereignty: Toward an agenda. Acton, ACT: Australian National University Press.

Koepke, J. L., & Robinson, D. G. (2018). Danger ahead: Risk assessment and the future of bail reform. *Washington Law Review, 93*. Retrieved from papers.ssrn.com/abstract=3041622.

Noble, S. U. (2018). *Algorithms of oppression: How search engines reinforce racism.* New York, N.Y.: NYU Press.

Oguamanam, C. (2019). *Indigenous data sovereignty: Retooling Indigenous resurgence for development.* Waterloo, CA: Centre for International Governance Innovation. Retreived from cigionline.org/publications/indigenous-data-sovereignty-retooling-indigenous-resurgence-development.

O'Neil, Cathy. (2016). *Weapons of math destruction: How big data increases inequality and threatens democracy.* New York, N.Y.: Crown Publishing.

Computation as Cultural Material

Harrell, D. F. (2007). *Cultural roots for computing: The case of African diasporic orature and computational narrative in the GRIOT system.* Proceedings of the 2007 Digital Arts and Culture Conference, Perth, Australia.

___. (2013). *Phantasmal media: An approach to imagination, computation, and expression.* Cambridge, MA: The MIT Press.

Jensen, C. B., & Blok, A. (2013). Techno-animism in Japan: Shinto cosmograms, actor-network theory, and the enabling powers of non-human agencies. *Theory, Culture & Society, 30*(2): 84–115.

Lewis, J. E. (2014). A better dance and better prayers: Systems, structures, and the future imaginary in Aboriginal new media. In S. Loft & K. Swanson, K. (Eds.), *Coded territories: Tracing Indigenous pathways in new media art* (pp. 48–77). Retrieved from deslibris.ca.

___. (2016). Preparations for a haunting: Notes towards an Indigenous future imaginary. In D. Barney, G. Coleman, C. Ross, J. Sterne, & T. Tembeck (Eds.), *The participatory condition in the digital age* (pp. 229–49). Minneapolis, MN: University of Minnesota Press.

Kitano, N. (2007). *Animism, rinri, modernization; The base of Japanese robotics.* Paper presented at the Institute of Electrical and Electronics Engineers (IEEE) International Conference on Robotics and Automation, Rome, Italy. Retrieved from roboethics.org/icra2007/contributions.html.

Computational Culture

Bleeker, J. (2001). The race for cyberspace: Information technology in the Black diaspora. *Science as Culture* 10(3), 353-374.

Bratton, B. H. (2016). *The stack: On software and sovereignty.* Cambridge, MA: The MIT Press.

Capurro, R. (2008). Information ethics for and from Africa. *Journal of the American Society for Information Science and Technology* 59(7), 1162-1170. doi.org/10.1002/asi.20850.

Finn, E. (2017). *What algorithms want: Imagination in the age of computing*. Cambridge, MA: The MIT Press.

Golumbia, D. (2009). *The cultural logic of computation*. Cambridge, MA: Harvard University Press.

In response to Bruce Sterling's 'Essay on the New Aesthetic' [Article]. (2012). Retrieved from creators. vice.com/en_us/article/eza9xa/in-response-to-bruce-sterlings-essay-on-the-new-aesthetic.

Mavhunga, C. C. (Ed.). (2017). *What do science, technology, and innovation mean from Africa?* Cambridge, MA: The MIT Press.

McCue, M. & Holmes, K. (2018). Myth and the making of AI. *Journal of Design and Science*. doi. org/10.21428/d3a0f14d.

McKelvey, F. (2018). *Internet daemons: Digital communications possessed*. Minneapolis, MN: University of Minnesota Press.

Sterling, B. (2012). An essay on the New Aesthetic. Retrieved from wired.com/2012/04/an-essay-on-the-new-aesthetic/.

Non-Humans, Things, Objects

Barad, K. (2007). *Meeting the universe halfway: Quantum physics and the entanglement of matter and meaning*. Durham, NC: Duke University Press.

Bennett, J. (2009). *Vibrant matter: A political ecology of things*. Durham, NC: Duke University Press.

Bogost, I. (2012). *Alien phenomenology, or what it's like to be a thing*. Minneapolis, MN: University of Minnesota Press.

Davis, H. [Sonic Acts]. (2016, February 28). *The queer futurity of plastic* [Video]. Retrieved from vimeo. com/158044006.

Harman, G. (2010, July 23). Brief SR/OOO tutorial [Blog post]. Retrieved from doctorzamalek2. wordpress.com/2010/07/23/brief-srooo-tutorial.

Harvey, G. (2005). *Animism: Respecting the living world*. New York, NY: Columbia University Press.

Haraway, D. (2016). *Staying with the trouble: Making kin in the Chthulucene*. Durham, NC: Duke University Press.

___. (1991). *Simians, cyborgs and women: The reinvention of nature*. New York, NY: Routledge.

Morton, T. (2013). *Realist magic: Objects, ontology, causality*. London: Open Humanities Press.

___. (n.d.). OOO [Blog post]. Retrieved from ecologywithoutnature.blogspot.com/p/ooo-for-beginners.html.

Reiss, D., Gabriel, P., Gershenfeld, N., & Cerf, V. (2013, February). The interspecies internet? An idea in progress [Video file]. Retrieved from ted.com/talks/diana_reiss_peter_gabriel_neil_gershenfeld_and_vint_cerf_the_interspecies_internet_an_idea_in_progress.

Science Fiction

Gibson, W. The Sprawl Trilogy:

Gibson, William. (1986). *Count Zero*. London, GB: Victor Gollancz, Ltd.

___. (1988). *Mona Lisa Overdrive*. London, GB: Victor Gollancz, Ltd.

___. (1984). *Neuromancer*. New York, NY: Ace Science Fiction Books.

Hausman, B. M. (2011). *Riding the Trail of Tears*. Lincoln, NE: University of Nebraska Press.

Kwaymullina, A. The Tribe Trilogy:

Kwaymullina, A. (2013). *The Disappearance of Ember Crow*. Newtown, N.S.W.: Walker Books.

___. (2015). *The Foretelling of Georgie Spider*. Newtown, N.S.W.: Walker Books.

___. (2012). *The Interrogation of Ashala Wolf*. Newtown, N.S.W.: Walker Books.

Roanhorse, R. (2017). Welcome to Your Authentic Indian Experience™. *Apex Magazine, 99*. Retrieved from apex-magazine.com/welcome-to-your-authentic-indian-experience.

Taylor, D. H. (2016). I Am...Am I. In *Take us to your chief: And other stories* (24-45). Madeira Park, CA: Douglas & McIntyre.

___. (2016). Mr .Gizmo. In *Take us to your chief: And other stories* (77-91). Madeira Park, CA: Douglas & McIntyre.

Manifestos/Guidelines/Declarations

Ochigame, R. (2019, December 20). The invention of "ethical AI": How big tech manipulates academia to avoid regulation. *The Intercept*. Retrieved from theintercept.com/2019/12/20/mit-ethical-ai-artificial-intelligence.

The European Commission Joint Research Centre. (2018). *Declaration of Cooperation on Artificial Intelligence*. Brussels, Belgium. Retrieved from ec.europa.eu/jrc/communities/en/node/1286/document/eu-declaration-cooperation-artificial-intelligence.

Montréal declaration on responsible AI development. (2018). Montréal, CA: Université de Montréal. Retrieved from montrealdeclaration-responsibleai.com/the-declaration/.

The Toronto Declaration: Protecting the rights to equality and non-discrimination in machine learning systems. (2018). Retrieved from accessnow.org/the-toronto-declaration-protecting-the-rights-to-equality-and-non-discrimination-in-machine-learning-systems/.

6.3

Participants' Biographies

Organizers

Prof. Jason Edward Lewis (*Cherokee, Hawaiian and Samoan*) is the University Research Chair in Computational Media and the Indigenous Future Imaginary, at Concordia University, Montreal, Canada. He directs the Initiative for Indigenous Futures and co-directs Aboriginal Territories in Cyberspace and the Skins Workshops on Aboriginal Storytelling and Video Game Design. Lewis' creative work has been recognized with the inaugural Robert Coover Award for Best Work of Electronic Literature, a Prix Ars Electronica Honorable Mention, several imagineNATIVE Best New Media awards and six solo exhibitions. He's the author of numerous chapters in collected editions covering Indigenous technology and digital media, mobile media, video game design, machinima and experimental pedagogy with Indigenous communities. Lewis has worked in a range of industrial technology research settings, including Interval Research, US West's Advanced Technology Group, the Institute for Research on Learning, and Arts Alliance Lab. Lewis was born and raised in northern California.

Angie Abdilla (*Trawlwoolway*) is the founder & CEO of Old Ways, New. Abdilla works across culture, research, strategy and technology, with Country (known as an entity) centring how Indigenous cultural knowledges inform service design and deep technology for both the public and private sectors. Her published research on Indigenous Knowledge Systems, Robotics and Artificial Intelligence was presented at the United Nations Permanent Forum on Indigenous Issues, where she continues this work to inform the rights of future technologies. Abdilla publicly presents and lectures on Human/Technology inter-Relations at the University of Technology Sydney. Abdilla is a Fellow of The Ethics Centre and holds a Bachelor of Arts in Communication.

Dr. ʻŌiwi Parker Jones (*Kanaka Maoli*) is a Research Fellow at the University of Oxford where he works on biological and artificial intelligence in the departments of Neuroscience and Engineering. In the 1980s, he was among the first children to be raised speaking Hawaiian in two generations. Later, as a graduate student, he worked on the adaptation of big data computing for the often fragmented corpora available in endangered languages—a research programme that he has continued to advance, for example by developing hybrid Deep Learning methods that contribute to the preservation and revitalisation of the Hawaiian language (e.g. Shillingford and Parker Jones 2018). As a postdoc, Dr. Parker Jones trained in systems neuroscience—with an emphasis on applications of machine learning to large-scale brain data. His current research is focused on Brain Computer Interfaces.

Dr. Noelani Arista (*Kanaka Maoli*), Researcher, Writer, Historian, is Associate Professor of Hawaiian and American History at the University of Hawai'i-Mānoa. Her research and writing focus on Hawaiian religious, legal, and intellectual history. Dr. Arista's current projects further the persistence of Hawaiian historical knowledge and Hawaiian language textual archives through multiple digital mediums including gaming. Dr. Arista is known for her work in developing new approaches and methods for writing Hawaiian history up from customary modes of keeping Hawaiian knowledge. Her work has also focused on precision in crafting historical contexts as an important first step in approaching the interpretation and translation of Hawaiian language sources. Her work in historiography, the training of Hawaiian intellectuals, as well as translation has prepared her for considering larger questions of cognition, and how artificial intelligence might be created and approached on Hawaiian terms. She mentors many students, instructing them in how to conduct research in Hawaiian language textual archives, and through online digital mediums. She was a contributing author to "Making Kin with Machines," an essay about Indigenous views on Artificial Intelligence, one of ten award winning essays in the MIT competition, *Resisting Reduction*. Her book *The Kingdom and the Republic: Sovereign Hawai'i and the Early United States* was published by PENN press in 2019. Her creative projects include the extensive facebook archive of mele, translation and photos that she wrote and compiled, *365 Days of Aloha*.

Suzanne Kite is an Oglála Lakȟóta performance artist, visual artist, and composer and a PhD candidate at Concordia University and Research Assistant for the Initiative for Indigenous Futures, and a 2019 Trudeau Scholar. Her research is concerned with contemporary Lakota epistemologies through research-creation, computational media, and performance practice. Recently, Kite has been developing a body interface for movement performances, carbon fiber sculptures, immersive video & sound installations.

Michelle Lee Brown is Euskaldun, Miarrtiz area (Côte des Basques) and German/German American, but raised on the lands and waters of the Wampanoag. As a PhD candidate, she studies Indigenous political praxis and futures through Indigenous designers' video games, graphic novels, and machinima at University of Hawai'i at Mānoa on the mokupuni of O'ahu in the Kona moku, part of the traditional and ongoing sovereign territories of the Kānaka Maoli. Brown has published peer-reviewed work on the Never Alone video game, a chapter on immersive media for Routledge's forthcoming *Handbook on Popular Culture and World Politics*, a chapter on Thunderbird Strike for *"The Women, They Hold the*

Ground": Indigenous Women's Digital Media in North America from University of Minnesota Press, and a comic in the recent *Relational Constellation* collection from MSU Press and Native Realities Press. She is currently working on a VR project and completing her dissertation *(Re)coding Survivance: Indigenous Media Science and Relation-Oriented Ontologies.*

Participants

Brent Barron is Director, Public Policy at CIFAR where he is responsible for engaging the policy community around cutting edge science. He played an important role in the development of the Pan-Canadian Artificial Intelligence Strategy, and now oversees CIFAR's AI & Society program, examining the social, ethical, legal, and economic effects of AI. Prior to this role, Barron held a variety of positions in the Ontario Public Service, most recently in the Ministry of Research, Innovation and Science. Brent holds a Master's in Public Policy from the University of Toronto, as well as a Bachelor's in Media Studies from Western University.

Scott Benesiinaabandan is an Anishinaabe intermedia artist that works primarily in digital media, including photography, video, audio, VR and installations. Scott has completed national and international residencies at Parramatta Artist Studios in Australia, Context Gallery in Derry, North of Ireland, and University Lethbridge/Royal Institute of Technology iAIR residency, Initiatives for Indigenous Futures, along with international collaborative projects in both the UK and Ireland. Scott is from Winnipeg and is currently based in Montreal, where he is completing a MFA in Studio Arts at Concordia.

Meredith Coleman received her BA (Hons) in English literature from the University of Winchester. She is an aspiring writer and has a deep-rooted interest in anthropology and sociology, but a lesser grasp of AI and technology studies. Coleman hopes that being involved in this project will help her to gain insight into a different area of academia—one that she have observed from a young age, through her family upbringing and overlaps with degree subjects.

Dr. Ashley Cordes (*Coquille*) is an Assistant Professor at the University of Utah in Indigenous Communication. Her research lies at the intersections of communication, digital media, and Indigenous studies and is attuned to issues of social power and decolonization. Recent work focuses on crypto and land-based currency as media, and on cultural appropriation in electronic dance music contexts. Cordes' work can be found in peer-reviewed journals including *Television & New Media* and *New Media & Society*. She has a professional background in multiplatform journalism and is currently a 2018-2019 American Philosophical Society Digital Knowledge Sharing Fellow, and Chair of the Culture and Education Committee of the Coquille Indian Tribe.

Kaipu (Kaipulaumakaniolono) Baker Hailing from the lush and cascading cliffs of the Koʻolau in the verdant ahupuaʻa of Kahaluʻu on the island of Oʻahu a Lua in the center of the Hawaiʻi archipelago, Kaipulaumakaniolono recognizes first and foremost the cloud banks that bud at the lofty peaks of those sacred cliffs. A graduate of the Kamehameha Schools Kapālama in 2016 and the University of Hawaiʻi at Mānoa in 2019 with bachelors in both English Literature and Hawaiian Language, he continues his studies in the MFA for Hawaiian Theatre program at UHM. His work and research focuses on excellence in Moʻolelo Kaʻao, traditional storytelling, and Mele, song and chant. Kaipu has worked as a tutor of Hawaiian language and appeared most notably in the productions of Kamapuaʻa (2006, 2007, 2008), Lāʻieikawai (2015), and as "Maui" in the Hawaiian language dubbing of Moana (2018). Kaipu practices indigenous futurity in the form of reshaping and remembering traditional narratives, i noho haku ai kanaka maoli i ka moʻolelo maoli o ia lāhui.

Dr. Melanie Cheung is an award-winning neurobiologist from Central North island tribe Ngāti Rangitihi. She is passionate about transforming therapeutic approaches to brain diseases, with less emphasis on drugs, more emphasis on structurally and functionally changing the brain through neuroplasticity-based technologies. Melanie's research is underpinned by a belief that that there is significant untapped knowledge and potential within Māori intellectual traditional that has the power to benefit humankind. Subsequently her work has involved intensive Māori community engagement (with elders and families with brain diseases) and development of decolonizing methodologies (incorporating Māori protocols into scientific and clinical practices).

Joel Davison is a Gadigal and Dunghutti man from Sydney Australia. Living culture through an active role in language revitalisation for the Gadigal language, he is also an avid technologist and works at the Commonwealth Bank of Australia as a Robotics Analyst.

Kūpono Duncan is a Native Hawaiian artist from Kailua, Oʻahu. His artwork primarily attempts to bridge motifs of the past with experiences in the present, using contemporary mediums. Kūpono has numerous years of experience as a muralist, contributing to pieces on display at the Hawaiʻi Convention Center, Bishop Museum, Sheraton Waikiki, Mokulēʻia, The Hawaiʻi Institute of Marine Biology on Moku o Loʻe, and various buildings around Honolulu. He strives continuously to perpetuate his culture through multimedia art.

Rebecca Finlay leads CIFAR's strategy to connect outstanding researchers with thought leaders who thrive on research insights relevant to the future of policy, business, health, and international development. She works with a team of knowledge mobilization experts who specialize in knowledge exchange, government relations, public policy, and innovation. In 2017, they launched CIFAR's AI & Society program that supports the examination of questions AI will pose for all aspects of society such as the economy, ethics, policymaking, philosophy, and the law. Her team also builds partnerships with governments across Canada and internationally. Prior to joining CIFAR, Finlay held leadership roles in research and civil society organizations including as Group Director, Public Affairs and Cancer Control for the Canadian Cancer Society and National Cancer Institute of Canada. She began her career in the private sector building strategic partnerships, including as First Vice President, Financial Institution and Partnership Marketing for Bank One International. Rebecca holds an M.Phil. in Social and Political Sciences from the University of Cambridge.

Sergio Garzon was born in Bogota, Colombia and lives and works in Honolulu, Hawai'i. His paintings and prints consist of abstract figurative narratives of his memories in Colombia focusing on people, culture and the politics of history. The visual contrast of his work comes from living in Colombia during a period of turmoil with Colombia's two predominant rebel groups, The Revolutionary Armed Forces of Colombia (FARC) and the National Liberation Army (ELN). His work employs sculpture, video, photography, printmaking, painting, performance and installation, often in unexpected combinations that traverse traditional practice boundaries. He is best at solving visual puzzles through the manipulation of natural bi-products of fire, earth and light.

D Fox Harrell, Ph.D., is Professor of Digital Media & Artificial Intelligence in the Comparative Media Studies Program and Computer Science and Artificial Intelligence Laboratory (CSAIL) at MIT. He is the director of the MIT Center for Advanced Virtuality. His research explores the relationship between imagination and computation. His research involves developing new forms of VR, computational narrative, videogaming for social impact, and related digital media forms based in computer science, cognitive science, and digital media arts. The National Science Foundation has recognized Harrell with an NSF CAREER Award for his project "Computing for Advanced Identity Representation." Dr. Harrell holds a Ph.D. in Computer Science and Cognitive Science from the University of California, San Diego. His other degrees include a Master's degree in Interactive Telecommunication from New York University, and a B.F.A. in Art (electronic and time-based media), B.S. in Logic and Computation from Carnegie Mellon University (each with highest honors). He has worked as an interactive television producer and as a game designer. His book *Phantasmal Media: An Approach to Imagination, Computation, and Expression* was published by the MIT Press.

Kekuhi Keali'ikanaka'oleohaililani (*Kanaka'ole 'Ohana-Pele Clan*) is an educator, scholar, dancer, musician, vocalist, composer, and powerful leader, as well as wife, mother, and daughter. She grew up on the slopes of the volcano Mauna A Wakea and Mauna Loa, in the daily influence of Kilauea, regarded as a family ancestor. Fluent in Hawaiian as well as English, educated in Hawaiian tradition and earning advanced degrees in Western universities, she defines what it means to be an Indigenous intellectual in a contemporary world. Through her visionary leadership, she engages Indigenous thought and knowledge to address today's issues through music, chant, and sharing of the spirit.

Megan Kelleher is embarking on her PhD as one of RMIT's Vice Chancellor's Indigenous Pre-Doctoral Fellows in the School of Media and Communication. The working title of her thesis is 'Blockchain, Black chains and the battle for systems sovereignty: mutual solutions for governance using Indigenous Knowledge (IK) systems and Indigenous-controlled protocols within the Blockchain'. The research seeks to explore the logical, structural or architectural synergies – or incompatibilities – between IK systems and Blockchain technologies, and the opportunities to embed IK approaches into second-wave automation. Grounded in her Barada/Baradha and Gabalbara/Kapalbara heritage, the research will be approached from an Indigenous standpoint, contributing to the field from an important Australian research perspective. Previous to RMIT Megan was at Creative Victoria in Indigenous Partnerships, and in the Department of Premier and Cabinet's Strategic Communication and Protocol Branch.

Maroussia Lévesque is an attorney and researcher with a background in interactive media. She consults for governments, private sectors, and NGOs about the legal and policy implications of emerging technologies. She was the Conceptual Lead at Obx Labs for Experimental Media during her B.F.A in Computation Arts at Concordia University, and researched IP issues at the Center for Genomics and Policy during her B.C.L./LL.B. law degrees from McGill. Maroussia was involved in the Quebec inquiry commission on the electronic surveillance of journalists, and drafted a foreign policy pertaining to AI and human rights for the Digital Inclusion Lab at Global Affairs Canada. She is a member of the Institute of Electrical and Electronics Engineers working group on algorithmic bias and speaks about law in digital spaces in contexts ranging from informal privacy workshops to international conferences and peer-reviewed journals.

Olin Lagon (*Kanaka Maoli*) is a serial social entrepreneur, innovator and community organizer, currently focused on clean energy. He founded multiple companies, nonprofits, and a foundation including one of the first crowdfunding companies which channeled $100 million to causes worldwide. He holds multiple patents and his designs have been adopted by Global 1000 companies and institutions like MIT. His service includes the U.S. Navy, the Peace Corps, and numerous nonprofits. He is a past Petra Fellow (Center for Community Change) and East West Center Fellow. Part Hawaiian and Filipino and raised in public housing, Lagon lives in Kalihi Valley with his wife and two young sons.

Dr. Jason Leigh is the Director of LAVA: the Laboratory for Advanced Visualization & applications, and Professor of Information & Computer Sciences at the University of Hawaiʻi at Mānoa. He is also Director Emeritus of the Electronic Visualization Lab and the Software Technologies Research Center at the University of Illinois at Chicago, where he was previously Professor of Computer Science and Affiliated Professor of Communications. In addition he was a Fellow of the Institute for Health Research and Policy at the University of Illinois at Chicago, and has held research appointments at Argonne National Laboratory, and the National Center for Supercomputing Applications. His research expertise includes big data visualization, virtual reality, high performance networking, and video game design. He is co-inventor of the CAVE2 Hybrid Reality Environment, and SAGE: Scalable Amplified Group Environment software, which has been licensed to Mechdyne Corporation and Vadiza Corporation, respectively. In 2010 he initiated a new multi-disciplinary area of research called Human Augmentics which refers to the study of technologies for expanding the capabilities and characteristics of humans. His research has also received numerous press from news media including the *AP News*, *The New York Times*, *Popular Science's Future Of*, *Nova ScienceNow*, *NSF Science Now*, *PBS*, and *Forbes*. Leigh also teaches classes in Software Design, Virtual Reality, Data Visualization and Video Game Design. In 2010 his video game design class enabled the University of Illinois at Chicago to be ranked among the top 50 video game programs in US and Canada.

Keoni Mahelona is currently building Te Reo Māori speech recognition tools including text to speech, speech to text, and measuring pronunciation. Mahelona's main roles are project management and web development, primarily for koreromaori.com and koreromaori.io. They also built the indigenous media platform tehiku.nz which serves as a digital Marae for Te Hiku Media and the five Iwi of Muriwhenua. Their key contribution is the Kaitiakitanga License which serves to guard Indigenous data and IP from misuse while aiming to create opportunities for the advancement of Indigenous peoples.

Caleb Moses (*Aotearoa Māori*) is a Data Scientist hailing from the Hokianga region in the far north of New Zealand. He has a Postgraduate Diploma in Pure Mathematics from the University of Auckland. His work focuses on machine learning, natural language processing, and automation. Moses is currently working with Te Reo Irirangi o te Hiku o te Ika on language technologies for Te Reo Māori, the language of the indigenous people of Aotearoa New Zealand.

Issac Nahuewai 'Ika'aka (Isaac) is a choice taro corm that comes from the rains that sound the metrosideros polymorpha flowers of Hilo. Educated at the University of Hawaii at Hilo with a B.A. in Hawaiian Studies and Anthropology, he is currently in the M.A. program studying Hawaiian Language and Literature. On top of being a student, he is also a part-time teacher at Ka Haka 'Ula o Ke'elikōlani, College of Hawaiian Language at UH Hilo and Ke Kula 'o Nāwahīokalani'ōpu'u Public Charter School. Outside of his roles in education, 'Ika'aka loves educating people through musical vibrations; he is a musical director for many bands around Hilo that spread conscious messages through reggae and jazz. He firmly believes that music can be an effective mode to revivify the value of ancestral knowledge and cultural identity in indigenous people.

Kari Noe is both a creative media and software developer originally from Kaua'i, now based in Honolulu, O'ahu. She has earned two bachelor's degrees, one in Computer Science and the other in Animation through the Academy of Creative Media at the University of Hawai'i at Mānoa. Currently she is a Graduate Research Assistant at the Laboratory for Advanced Visualization and Applications (LAVA), pursuing a master's degree in Computer Science. Kari has worked on various projects from creating her own animated film, Kai and Honua, to collaborating on a virtual reality Hawaiian navigation application named Kilo Hōkū. She specializes in virtual reality and augmented reality research for cultural preservation and is currently working on her thesis, with the working title: Digitizing Detours, Mapping Hawaiian Knowledge in Virtual Reality.

Danielle Olson is a PhD student in Electrical Engineering & Computer Science at MIT and works as a Research Assistant in the Imagination, Computation, and Expressions (ICE) Lab within the MIT Computer Science and Artificial Intelligence Laboratory. Olson's research seeks to develop theories and technologies to advance an understanding of embodied identity expression in virtual reality (VR) narratives to reflect the nuance of real-world human interaction. Olson earned her B.S. in Computer Science & Engineering from MIT in 2014, and her S.M. in Electrical Engineering and Computer Science from MIT in 2019. While at MIT, Olson founded Gique Corporation, an educational nonprofit

501(c)(3) that exists to inspire and educate youth in STEAM. Following her graduation from MIT, Danielle worked as a Program Manager at the Microsoft New England Research & Development Center from 2014-2016. Danielle also previously worked as Summer Program Coordinator for the MIT Online Science, Technology, and Engineering Community (MOSTEC) in the summer of 2016, prior to returning to MIT as a graduate student.

Archer Pechawis (*Plains Cree*) is a performance, theatre and new media artist, filmmaker, writer, curator and educator born in Alert Bay, BC. He has been a practicing artist since 1984 with a particular interest in the intersection of Plains Cree culture and digital technology, merging "traditional" objects such as hand drums with digital video and audio sampling. His work has been exhibited across Canada, internationally in Paris France and Moscow Russia, and featured in publications such as Fuse Magazine and Canadian Theatre Review. Archer has been the recipient of many Canada Council, British Columbia and Ontario Arts Council awards, and won the Best New Media Award at the 2007 imagineNATIVE Film + Media Arts Festival and Best Experimental Short at imagineNATIVE in 2009. Archer has worked extensively with Native youth since the start of his art practice, originally teaching juggling and theatre, and now digital media and performance. He is currently a member of the Indigenous Routes collective, teaching video game development to Native girls: www.indigenousroutes.ca. Of Cree and European ancestry, he is a member of Mistawasis First Nation, Saskatchewan.

Caroline Running Wolf (*Crow Nation*), née Old Coyote, is an enrolled member of the Apsáalooke Nation (Crow) in Montana, with a Swabian (German) mother and also Pikuni, Oglala, and Ho-Chunk heritage. As the daughter of nomadic parents, she grew up between USA, Canada, and Germany. Thanks to her genuine interest in people and their stories, she is a multilingual Cultural Acclimation Artist dedicated to supporting Indigenous language and culture vitality. After working for over 15 years as a professional nerd herder and business consultant in various fields, Running Wolf co-founded a nonprofit, Buffalo Tongue, with her husband, Michael Running Wolf. Together they create virtual and augmented reality experiences to advocate for Native American voices, languages, and cultures. Running Wolf has a Master's degree in Native American Studies from Montana State University in Bozeman, Montana. She is currently pursuing her PhD in Anthropology at the University of British Columbia in Vancouver, Canada.

Michael Running Wolf (*Northern Cheyenne*) was raised in a rural village in Montana with intermittent water and electricity. Naturally, he now has a Master's of Science in Computer Science. Though he is a published poet, he is a computer nerd at heart. His lifelong goal is to pursue endangered indigenous language revitalization using Augmented Reality and Virtual Reality (AR/VR) technology. He was raised with a grandmother who only spoke his tribal language, Cheyenne, which like many other indigenous languages, is near extinction. By leveraging his advanced degree and technical skills, Running Wolf hopes to strengthen the ecology of thought represented by indigenous languages through immersive technology.

Skawennati makes art that addresses history, the future, and change from her perspective as an urban Kanien'kehá:ka woman and as a cyberpunk avatar. Her work has been widely presented in both group exhibitions and solo shows and is included in public and private collections, such as the National Gallery of Canada and the Musée d'art contemporain de Montréal. Born in Kahnawà:ke Mohawk Territory, Skawennati graduated with a BFA from Concordia University in Montreal, where she is based. She is Co-Director of Aboriginal Territories in Cyberspace (AbTeC), a research-creation network of artists and academics who investigate and create Indigenous virtual environments. Their Skins workshops in Aboriginal Storytelling and Experimental Digital Media are aimed at empowering youth. In 2015 they launched IIF, the Initiative for Indigenous Futures.

Tyson Seto-Mook received his BS in Electrical Engineering and is currently pursuing a MS in Computer Science from the University of Hawai'i at Mānoa.

Dr. Hēmi Whaanga (*Ngāti Kahungunu, Ngāi Tahu, Ngāti Mamoe, Waitaha*) is an associate professor in Te Pua Wānanga ki te Ao (The Faculty of Māori and Indigenous Studies) at Te Whare Wānanga o Waikato (University of Waikato). Whaanga has worked as a project leader and researcher on a range of projects centred on the revitalisation, protection, distribution, and development of Mātauranga and te reo Māori in a digital world. He incorporates multi-method techniques and methodologies to analyse and develop new Mātauranga in a range of linguistic, cultural, and digital contexts including the design of ethical platforms for digitally managing and distributing Mātauranga, oral traditions, Māori ecological knowledge, ecological taxonomies, and naming protocols, Māori astronomical knowledge and kaitiakitanga. He affiliates to Ngāti Kahungunu through his father, and Ngāi Tahu, Ngāti Mamoe, and Waitaha through his mother.

6.4
Workshop Schedules

March 1–2, 2019

Indigenous Protocol and Artificial Intelligence

Workshop 1

Honolulu, Oahu, Hawai'i
www.indigenous-ai.net

The Indigenous Protocol and Artificial Intelligence (A.I.) Workshops will focus on how to advance the theory and practice of next-level A.I. from Indigenous perspectives.

We will consider the following questions:

- From an Indigenous perspective, what should our relationship with A.I. be?

- How can Indigenous epistemologies and ontologies contribute to the global conversation regarding society and A.I.?

- How do we broaden discussions regarding the role of technology in society beyond the largely culturally homogenous research labs and Silicon Valley startup culture?

- How do we imagine a future with A.I. that contributes to the flourishing of all humans and non-humans?

Global Organizers

Jason Edward Lewis

Angie Abdilla

ʻŌiwi Parker Jones

RA: Suzanne Kite

Local Organizers

Noelani Arista

RA: Michelle Brown

Flights + Hotels

Brent Barron

Jacqui Sullivan

Venues

LAVA Lab
Keller Hall 102, 2550 Correa Rd
University of Hawaiʻi at Mānoa
Honolulu, HI 96822

Ka Waiwai
#100, 1110 University Ave
Honolulu, HI 96825

Lincoln Hall
1821 East-West Rd
University of Hawaiʻi at Mānoa
Honolulu, HI 96848

Ala Moana Hotel
410 Atkinson Dr
Honolulu, HI 96814

Day 1

Friday
March 1

8:30am	Light Breakfast
9:00am	Welcome & Introductions
10:30am	Break
11:00am	Protecting Indigenous Cultural Knowledge 1
12:00pm	Lunch
1:30pm	Futuring Exercise
2:30pm	Discuss Blogging Questions
3:30pm	Break
4:00pm	Construct Themes
5:00pm	Protecting Indigenous Cultural Knowledge 2
5:30pm	Closing
6:00pm	Dinner
7:00pm	ʻAwa & ʻAi Ka Waiwai public event

Location
Ka Waiwai

#100, 1110 University Avenue
Honolulu, HI 96825

Day 2

Saturday
March 2

8:30am	**Light Breakfast**
9:00am	**Review of Day 1**
9:30am	**Thematic Breakout Groups** Break as needed
12:00pm	**Lunch**
1:30pm	**Share Breakout Results**
3:00pm	**Break**
3:30pm	**Next Steps**
5:00pm	**Protecting Indigenous Cultural Knowledge 3**
5:30pm	**Closing**
7:00pm	**Dinner**

Location
LAVA Lab

Keller Hall 102, 2550 Correa Rd
University of Hawai'i at Mānoa
Honolulu, HI 96822

The Indigenous Protocol and Artificial Intelligence (A.I.) Workshops will develop new conceptual and practical approaches to building the next generation of A.I. systems.

We will consider the following questions:

- From an Indigenous perspective, what should our relationship with A.I. be?

- How can Indigenous epistemologies and ontologies contribute to the global conversation regarding society and A.I.?

- How do we broaden discussions regarding the role of technology in society beyond the largely culturally homogenous research labs and Silicon Valley startup culture?

- How do we imagine a future with A.I. that contributes to the flourishing of all humans and non-humans?

Sunday **May 26**	1:00pm	Welcome
	1:00pm—5:00pm	Review and Organization
	6:00pm	Dinner
Monday / Tuesday **May 27 / 28**	9:00am—4:00pm	Writing
	4:00pm—5:00pm	Group Review
	6:00pm	Dinner
Wednesday **May 29**	9:00am—12:00pm	Writing
	12:00pm	Group Outing
Thursday / Friday **May 30 / May 31**	9:00am—4:00pm	Writing
	4:00pm—5:00pm	Group Review
	6:00pm	Dinner
Saturday **June 1**	10:00am—1:00pm	Reviewing and Planning
	4:00pm	Open Invite BBQ

Organizers

Jason Edward Lewis

Noelani Arista

Suzanne Kite

Michelle Brown

Series Co-organizers

Jason Edward Lewis

Angie Abdilla

'Ōiwi Parker Jones

Founding Organizers

 Initiative for Indigenous Futures

 OLD WAYS, NEW

CIFAR

Support

Department of History +
College of Arts and Sciences
University of Hawai'i at Mānoa

 Social Sciences and Humanities Research Council of Canada   Conseil de recherches en sciences humaines du Canada
 Canada

 CONCORDIA RESEARCH CHAIR IN COMPUTATIONAL MEDIA & THE INDIGENOUS FUTURE IMAGINARY

SECTION 7
Acknowledgements

7.0

Acknowledgements

The organizers of the Indigenous Protocol and Artificial Intelligence Workshops would like to acknowledge the Canadian Institute for Advanced Research (CIFAR) for providing core funding through its the Pan-Canadian AI Strategy. Our main CIFAR liaison, Brent Barron, was a fruitful collaborator and tireless champion who worked extensively with us to craft workshops that were welcoming of Indigenous bodies and knowledges. We also wish to thank Jacqui Sullivan for the assistance on logistics she provided, and Rebecca Finlay for joining us alongside Brent in the first workshop.

We would also like to acknowledge the following for contributing their time, good minds and/or other resources to make the events a success:

The Initiative for Indigenous Futures and Old Ways, New for providing personnel and resources for the workshop organization and contributing substantial funding.

The Social Sciences and Humanities Research Council of Canada and the Concordia University Research Chair in Computational Media and the Indigenous Future Imaginary for providing additional funding and support.

Ty Kawika Tengan and his assistants, Kaipulaumakaniolono Baker and Isaac ʻIkaʻaka Nāhuewai, for welcoming us to Hawaiian territory on the first day of the first workshop.

Dr. Jason Leigh at the University of Hawaiʻi at Mānoa for graciously offering the use of his LAVA lab as well as facilitating the use of the Hawaiian Data Science lab space to host the second day of the first workshop.

Matt Lampert for sharing his substantial reference list on Indigenous knowledge frameworks and technology in African, South American and Asian contexts.

Position Paper Team

Editor: Jason Edward Lewis

Managing Editor: Mikhel Proulx

Editorial Support: Anastasia Erickson

Assistant Editors: Suzanne Kite and Michelle Lee Brown

Editorial Advisors: Dr. Hēmi Whaanga, Dr. Melanie Cheung, and Dr. Noelani Arista

CIFAR

This work was supported by CIFAR through the Pan-Canadian AI Strategy. Learn more about CIFAR at cifar.ca.

About CIFAR

CIFAR is a Canadian-based global charitable organization that convenes extraordinary minds to address the most important questions facing science and humanity.

About the AI & Society program

The AI & Society program is the fourth pillar of the CIFAR Pan-Canadian AI Strategy, a $125-million investment from the Government of Canada, with the goal of supporting Canada's leadership in machine learning and training. It is also supported by Facebook and the RBC Foundation. The AI & Society workshops are led by CIFAR, in partnership with the Centre national de la recherche scientifique (CNRS) and UK Research and Innovation (UKRI).

Document designed by Aimee Wood.

Gathered at Waiwai. Image by Sergio Garzon, 2019.